T0181972

The Cheeses of Italy: Science and Technology

Marco Gobbetti • Erasmo Neviani • Patrick Fox

The Cheeses of Italy: Science and Technology

With Contribution by Gian Maria Varanini

 Springer

Marco Gobbetti
Faculty of Science and Technology
Free University of Bolzano
Bolzano, Italy

Erasmo Neviani
Food and Drug Department
University of Parma
Parma, Italy

Patrick Fox
School of Food and Nutritional Sciences
University College
Cork, Ireland

ISBN 978-3-030-07877-5   ISBN 978-3-319-89854-4   (eBook)
https://doi.org/10.1007/978-3-319-89854-4

© Springer International Publishing AG, part of Springer Nature 2018
Softcover reprint of the Hardcover 1st edition 2018
This work is subject to copyright. All rights are reserved by the Publisher, whether the whole or part of the material is concerned, specifically the rights of translation, reprinting, reuse of illustrations, recitation, broadcasting, reproduction on microfilms or in any other physical way, and transmission or information storage and retrieval, electronic adaptation, computer software, or by similar or dissimilar methodology now known or hereafter developed.
The use of general descriptive names, registered names, trademarks, service marks, etc. in this publication does not imply, even in the absence of a specific statement, that such names are exempt from the relevant protective laws and regulations and therefore free for general use.
The publisher, the authors and the editors are safe to assume that the advice and information in this book are believed to be true and accurate at the date of publication. Neither the publisher nor the authors or the editors give a warranty, express or implied, with respect to the material contained herein or for any errors or omissions that may have been made. The publisher remains neutral with regard to jurisdictional claims in published maps and institutional affiliations.

Printed on acid-free paper

This Springer imprint is published by the registered company Springer International Publishing AG part of Springer Nature.
The registered company address is: Gewerbestrasse 11, 6330 Cham, Switzerland

M.G. dedicates this book to Camilla, as a small gift to fulfil the realization of her dreams
E.N. dedicates this book to Iole and Umberto, wishing them a life full of emotion and beauty

Preface

Many aspects of the Italian cheese industry are unique. Cheesemaking in Italy has a very long history, pre-dating Roman civilization and was well developed at the time of the Roman Empire. A number of Classical Roman writers, especially Columella, wrote in considerable depth on cheese technology. Subsequently (fifteenth century), the treatise *Summa lacticiniorum*, by Pantaleone da Confienza was one of best examples of a cheese encyclopedia, describing in detail Italian cheeses from different populations and territories.

A higher percentage (ca. 80%) milk from Italy is used for cheesemaking than from any other country. The Italian cheese industry is well organized and largely controlled by consortia, which supervise the technology and marketing of the particular varieties. The milk of four species, cow, sheep, goat, and water buffalo, is used for cheesemaking. About 450 varieties/variants of cheese are produced in Italy, several of which have Protected Denomination of Origin (PDO) status, and some are famous and copied around the world. Italy has one of the highest per capita consumptions of cheese: it is consumed as a table cheese, as an hors d'oeuvre, as a component of dishes, and as a sauce or flavoring. There is a substantial body of literature on Italian cheese, but most of it is descriptive. In this book, we describe the history of Italian cheesemaking, the technology of 59 varieties, the microbiology and biochemistry of the major varieties, and the principal distinguishing features of the major Italian cheeses.

Bolzano, Italy   Marco Gobbetti
Parma, Italy   Erasmo Neviani
Cork, Ireland   Patrick Fox
Verona, Italy   Gian Maria Varanini

Contents

About the Authors

Patrick F. Fox is Emeritus Professor of Food Chemistry at University College, Cork, Ireland. His teaching and research interests are on various aspects of dairy chemistry, especially the production and ripening of cheese, the heat stability of milk, the physico-chemical properties of milk proteins, and dairy enzymology. He is the author or co-author of around 600 peer-reviewed publications, author or editor of approximately 30 books on dairy chemistry, and one of the founding editors of the *International Dairy Journal* in 1990.

Marco Gobbetti has been a full professor of food microbiology since 2000. Currently, he is working at the Free University of Bolzano, Italy. His research, teaching, and outreach focuses are on food fermentations and lactic acid bacteria physiology and biochemistry. He has published articles and books on several fermented foods, using lactic acid bacteria as starters for food manufacturing, and on driving complex food ecosystems through the application of omic technologies. To date, he is author of more than 300 peer-reviewed articles, having an h-index of 60, with more than 12,000 citations. He is a member of the editorial board of various international journals. He holds a number of representative positions at national and international levels dealing with biotechnology and the life sciences.

Erasmo Neviani has been a full professor of agricultural microbiology since 2002 at the Food and Drug Department of the University of Parma. Formerly, he was Dean of the Faculty of Agriculture at the University of Parma (2009–2012), Head of the Microbiology and Enzymology Department of Dairy Research Institute (Lodi, Italy) of the Italian Ministry of Agriculture (1996–2002), and President of the SIMTREA, the Italian Society of Agricultural, Food and Environment (2013–2015).

His main research areas concern food microbiology, with specific expertise in dairy microbiology. The main part of his research work was dedicated to the study of lactic acid bacteria used in dairy fermentations, particularly with regard to their technological performance. His research activity can be summarized as follows: (1) the microbiology and technology of dairy products; (2) the molecular microbiology and biotechnology of lactic acid bacteria; (3) population dynamics and the behavior of lactic acid bacteria in food fermentation; (4) microbial interactions in natural food ecosystems; (5) microbial enzymatic activities and cheese ripening; (6) food safety and processing; and (7) the human intestinal microbiome in response to dietary habits, diseases, and food intolerance. He has authored approximately 300 publications that relate to the microbiology of dairy food, most of them (ca. 150) in international journals.

He is author of the book "Microbiologia e Technologia Lattiero-Casearia, Qualità e Sicurezza" 2006, Ed Tecniche Nuove (Milan) ISBN 8848118178". He is co-editor of an international book: Randazzo C. L., Caggia C., Neviani E. (2013): Cheese Ripening: Quality, Safety and Health Aspects. Nova Science Publishers. Hauppauge NY, USA. ISBN 978-1-62417.

He has also published various chapters in internationally and nationally published books.

Chapter 1
The Origins of Cheesemaking

1.1 Introduction

Cheese is the generic name for a group of fermented milk-based food products, produced in a wide range of flavors and forms throughout the world. The primary objective of cheesemaking is to conserve the principal constituents of milk, but cheese has evolved to become a highly nutritious food with epicurean qualities. Sandine and Elliker (1970) suggested that there are more than 1000 varieties of cheese. Walter and Hargrove (1972) described more than 400 varieties and listed the names of a further 400, while Burkhalter (1981) classified 510 varieties (although some are listed more than once). Harbutt (2002, 2009) described 750 cheese varieties, with photographs, and several cheese-based recipes. Barthilemy and Sperat-Czar (2001) described, with photographs, 1200 varieties. A list of 450 Italian cheeses, some with photographs, is given by Rubino et al. (2005) in Wikipedia, but many of these are variants; there are probably 40 distinct varieties of Italian cheese. Thirty-one varieties of Italian cheese were described, with photographs, by Jerry Finzi in Grand Voyage Italy in Wikipedia.

Several attempts have been made to classify cheese varieties into meaningful groups, see Chap. 4. The most common criterion for classification is texture (very hard, hard, semihard, semisoft, soft), which is related mainly to the moisture content of the cheese. Various attempts have been made to improve the classification on this basis, for example, by including the milk-producing species, moisture-to-protein ratio, method of coagulation, cooking temperature, type of secondary microbiota (e.g., blue or white mold), *Propionibacterium* or surface bacterial smear. However, no classification scheme developed to date is completely satisfactory; the inclusion of chemical indices of ripening would be useful.

© Springer International Publishing AG, part of Springer Nature 2018
M. Gobbetti et al., *The cheeses of Italy: Science and Technology*,
https://doi.org/10.1007/978-3-319-89854-4_1

1.2   Initiation of Cheesemaking

Cheese evolved in the "Fertile Crescent" (e.g., from the Tigris and Euphrates rivers), through what is now southern Turkey to the Mediterranean coast, some 8000 years ago. The so-called "Agricultural Revolution" occurred in this region with the domestication of plants and animals. Presumably, humans soon recognized the nutritive value of milk produced by domesticated animals and shared the mother's milk with her offspring. Goats and sheep, which are gregarious and docile, were the first dairy animals domesticated, but cattle have become the dominant dairy species in most parts of the world (ca. 85% of the global supply of milk is obtained from cattle, mainly *Bos taurus*). The water buffalo is an important dairy animal in India, Pakistan, Italy, and to a lesser extent elsewhere. A brief overview of the evolution of cheese is presented in this chapter (see also Fox and McSweeney 2004).

Milk is also a rich source of nutrients for bacteria, which contaminate the milk, some species of which utilize milk sugar, lactose, as a source of energy, producing lactic acid. When sufficient acid has been produced, the principal proteins of milk, the caseins, coagulate (e.g., at their isoelectric points, ca. pH 4.6), to form a gel in which the fat is occluded. The adventitious microbiota of milk is still the cause of acidification for many varieties of cheese, but cultures of lactic acid bacteria are used for the principal varieties.

When an acid-coagulated milk gel is broken, it separates into curds and whey; the acid whey is a pleasant drink for immediate consumption, while the curds may be consumed fresh or stored for future use. Whey was long considered to have medicinal benefits (Hoffmann 1761). The shelf life of the curds can be extended by dehydration and/or by adding NaCl; heavily salted cheese varieties are still widespread throughout the Middle East, and small quantities of a number of dehydrated cheeses are produced in North Africa and the Middle East, e.g., Tikammart and Aoules (Algeria), Djamid (Jordon), Ekt (Saudi Arabia), and Madraffarah (Syria) (Phelan et al. 1993).

1.3   Coagulation of Milk

One of the principal families of cheese, the acid-coagulated cheeses, modern members of which include cottage cheese, cream cheese, and quarg, originated in this way. Lactic acid, produced in situ, was probably the original milk coagulant, but an alternative mechanism was also recognized from an early date. Many proteolytic enzymes may modify the casein system of milk, causing it to coagulate under certain circumstances. Enzymes capable of causing this transformation are widespread in nature, e.g., bacteria, molds, plants, and animal tissues, but an obvious source would have been animal stomachs. Probably, it was observed that the stomach of young mammals after slaughter contained curds, especially if the animals had suckled shortly before slaughter; curds would also have been observed in the vomit of human infants. Before the development of pottery (ca. 5000 BC), storage of milk in bags made from animal skins was probably common (as it still is in many countries); stomachs of slaughtered animals provided ready-made, easily sealed containers. Under such circumstances, milk would extract enzymes (chymosin and some pepsin) from the stomach tissue, leading to its coagulation during storage. The properties of rennet-coagulated curds are very different from those produced by isoelectric (acid) precipitation, e.g., they have better syneretic properties, which makes it possible to produce low-moisture cheese curd. Rennet-coagulated curds, therefore, may be converted to a more stable product than acid curds, and rennet coagulation has become predominant in cheese manufacture, being exploited for ca. 75% of total world production.

Although animal rennets were used from early times, rennets produced from a range of plant species (e.g., fig and thistle) also appear to have been common in ancient times. However, plant rennets are not suitable for the manufacture of long-ripened cheese varieties, and gastric proteinases from young animals became the standard rennets until a shortage of supply made it necessary to introduce "rennet substitutes," around 1950.

While the coagulation of milk by the in situ production of lactic acid was, presumably, accidental, the use of rennets to coagulate milk was intentional. Indeed, it was quite an ingenious "invention." If the conversion of milk to cheese by the use of rennets was discovered today, it would be hailed as a major biotechnological discovery.

1.4 Cheese in Biblical Times

The advantages accruing from the ability to convert the principal constituents of milk to cheese would have been apparent from the viewpoints of storage stability, ease of transport and as a means of diversifying. The human diet and cheese manufacture became well established in the ancient civilizations of the Middle East, Egypt, Greece, and Rome. There are numerous references to cheese and other foods in the Bible (MacAlistair 1904). Milk and dairy products were important in the diet of peoples of the Near East during Biblical times; Palestine was "a land flowing with milk and honey" (*Exodus* **3**.8). Animals used for milk production during Biblical times included goats (e.g., *Proverbs* **27**.27), sheep (e.g., *Deuteronomy* **14**.4), and possibly camels (*Genesis* **32**.15). Bovine milk is rarely specified in the Old Testament, presumably because of the unsuitability of the terrain of Palestine for cow pasture; ancient cattle were larger and less docile than modern breeds. In addition to milk, other foods of dairy origin mentioned in the Bible include curds (perhaps fermented milk: *Genesis* **18**.8; *Isaiah* **7**.22) and butter (*Psalms* **55**.21). There are several references in the Old Testament to cheese, e.g., Job (1520 BC, where Job remarks to God "did Thou not pour me out like milk and curdle me like cheese"; *Job* **10**.10) and Samuel (1170–1017 BC, as a delicacy sent by Jesse to his sons (*I Samuel* **17**.18) and as a gift presented to David (*II Samuel* **17**.29).

Cheese is represented in the tomb art of Ancient Egypt and in Greek literature. Vegetable rennets are mentioned by Homer (about eighth century BC) who implies the use of fig rennet in the *Iliad* ("… as when fig juice is added to white milk and rapidly coagulates, and the milk quickly curdles as it is stirred, so speedy was his healing of raging Ares" *Iliad* **5**) and describes the Cyclops, Polyphemus, making ewe's milk cheese in the *Odyssey* (Book **9**) using well-made dairy vessels and pails swimming with whey (see also Chap. 2). Other Greek authors who mentioned cheese include Herodotus (484–408 BC), who referred to "Scythian cheese," and Aristotle (384–322 BC), who noted that Phrygian cheese was made from the milk of mares and asses. Cheese was prescribed in the diet for Spartan wrestlers in training.

1.5   Cheese in Classical Rome

Kindstedt (2012) provides a substantial account of the importance of cheese in Classical Greece; it was offered as a sacrifice to the Gods, used as a food at Symposia (drinking parties) and as part of the daily diet, either as a staple food or as a relish. Cheesemaking seems to have been important in Sicily, then part of *Magna Graecia*, since the fourth century BC, and Sicilian cheeses (pecorino and caprino types) were highly regarded and important exports to Greece (Kindstedt 2012). Little is known about the manufacture and nature of Greek and Sicilian cheeses.

Tuscany, on the central west coast of Italy, developed nearly a thousand years before Rome; it reached its peak about the sixth century BC and was subsumed by Rome in the first century BC. One of the characteristic Italian cheeses, Ricotta, originated in Tuscany. Ricotta was produced originally by heat-acid coagulation of whole or partly skimmed milk but is now produced from a mixture of milk (10–20%) and rennet whey; it is consumed fresh. The dairy animals in Tuscany were sheep and goats, and transhumance was normal. From the seventh century BC, rennet-coagulated aged pecorino and caprino cheeses were also produced in Tuscany (Kindstedt 2012).

Cheese manufacture was well established in the Roman Empire and was a standard item in the rations issued to Roman soldiers. Cheese must have been popular with Roman civilians also and demand exceeded supply, forcing an emperor, Diocletian (284–305 AD), to fix, in 301, a maximum price for cheese. Many Roman writers (e.g., Cato the Elder, 234–149 BC; Varro, (ca. 116–27 BC; Columella, 4-70 AD; Pliny the Elder, 2379 AD; and Palladius Rutilius Taurus Aemilenus, 400–470 AD) described cheese manufacture and quality and the culinary uses of cheese. Pliny mentioned cheese in his encyclopedia, *Historia Naturalis* (Book 28) and described its uses in the diet and in medicinal applications. Varro (in *De Agricultura* **2**.3–**2**.6) distinguished between "soft and new cheese" and that which is "old and dry" and reported spring–summer as the Roman cheesemaking season. He briefly described cheese manufacture. The milk was coagulated by rennet, a piece of

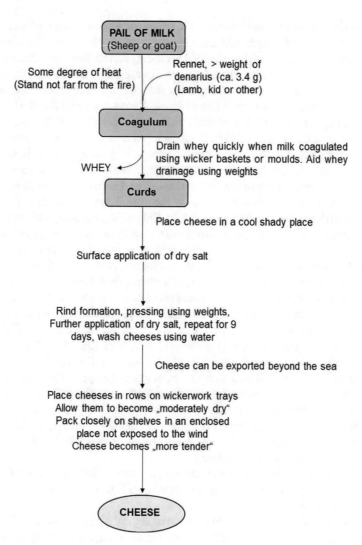

Fig. 1.1 Flow diagram for the manufacture of a type of Roman cheese based on the description of Columella (Adapted from De Re Rustica, 7.8.1–7.8.7)

stomach tissue from a hare or kid (in preference to that from a lamb). Fig latex or vinegar could be used as alternatives to rennet.

The most complete ancient description of Roman cheesemaking is that of Lucius Junius Moderatus Columella, a Roman soldier and author from Gades (modern Cadiz), in his treatise on agriculture, *De Re Rustica* (ca. AD 50). A manufacturing procedure for Roman cheese, based on the description of Columella, is given in Fig. 1.1, which includes many observations and practices familiar to modern cheesemakers. He recommended that the (raw) milk be held at "some degree of heat" but warns against overheating (e.g., by placing the pail on the flames of a fire). Columella distinguished between cheese with a "thin consistency" (soft?), which must be sold quickly while it is still fresh and retains its moisture, and that with a "rich and thick consistency" (hard?), which may be held for a long period. Since the concept of pH and the existence of bacteria were unknown in antiquity, no mention is made of starters. The cheese curd was acidified using the adventitious microorganisms of the raw milk. Columella discussed different types of rennet in some detail. He recommended coagulation using rennet from lamb or kid but reported the use of plant rennets also, e.g., flowers of certain thistles (perhaps *Cynara cardunculus*), the seeds of safflower (*Carthamus tinctorius*), or sap from the fig tree. Columella recommended that the smallest amount of rennet possible should be used to ensure high-quality cheese. Whey drainage was through wicker baskets, perhaps analogous to the drainage of whey through molds in the manufacture of certain soft cheeses (e.g., Camembert). No mention was made by Columella of cooking the curds–whey mixture prior to whey drainage; moisture control seems to have been by pressing the curds during whey drainage or pressing the cheese after salting. Salting involved the repeated application of dry salt to the cheese surface (which is still practiced, e.g., in the manufacture of blue cheese), which encouraged further loss of moisture ("acid liquid"). Columella also mentioned brine salting as a method of "hardening" cheese. The cheeses were washed with water, allowed to form a rind and placed on shelves in an enclosed place "so that the cheese may remain *more tender*." Columella also discussed defects which may occur in cheese, including being "full of holes" (mechanical openings), too salty or too dry. According to Columella, cheeses were flavored with herbs and colored with smoke, practices that persist to a certain extent today. He also described briefly the manufacture of "hand-pressed" (*manu pressum*) cheese in which hot water is poured over the curds which are then shaped by hand, a practice perhaps related to the kneading and stretching steps for *pasta filata* varieties. Thus, cheesemaking practice appears to have changed little from the time of Columella until the nineteenth century.

In his work, *Opus Agriculturae*, Palladius provided a brief description of cheesemaking, generally similar to, but less detailed than that of Columella, with whose work he was familiar.

Considering the long history of cheesemaking in Italy, it is not surprising that Italy is, arguably, the leading cheesemaking region in the world. A higher percentage, ca. 70%, of milk is converted into cheese in Italy than in any other country. The consumption of cheese in Italy is about 22 kg per caput, which is the fourth highest in the world; cheese is used in a wide range of dishes: antipasto, soups, sauces, bread, pizza, spaghetti dishes, lasagne, rice dishes, fish dishes, etc. Bovine, water buffalo, ovine, and caprine milk is used in substantial amounts, and some major and unique cheese varieties are produced, e.g., Parmigiano Reggiano, Grana Padano, Mozzarella di Bufala Campana, Ricotta, several Pecorino cheeses, Gorgonzola, and Taleggio.

Kindstedt (2012) traces the development and spread of cheese throughout the Middle East and into Europe: Mesopotamia (6000–4000 BC), Egypt (about 5000 BC), Indus valley (from 3000 BC), Hittite Empire, Anatolia (from 2000 BC), Minoan civilization, Crete (from 2000 BC), Mycenaean civilization, Greek mainland (from about 1500 BC) and into Europe. In many of these civilizations, cheese and butter were high-value products, given as gifts to the Gods, e.g., the Goddess Inanna, the Goddess of fertility and erotic love in the Mesopotamian city-state of Uruk.

A large area north of the Alps, from Hungary, through Austria, Switzerland, and into France, was inhabited by Celtic peoples. The Celts had a strong dairying tradition, mainly cattle. According to Kindstedt (2012), the Celts were famous for their cheeses, much of which was exported to Rome via Marseilles (Massalia). Kindstedt (2012) speculated that the Celts made large hard cheeses.

Although there were significant imports of cheese from several sources, especially from Greece, Sicily, Tuscany, and the Celts into Rome, it is probably true that the modern cheese industry emerged through the Roman Empire, which continued and expanded cheesemaking.

The great migrations of peoples throughout Europe immediately before and after the fall of the Western Roman Empire must have promoted the further spread of cheese manufacture, as did the Crusaders and other pilgrims of the Middle Ages. Probably, the most important agents contributing to the development of cheese technology and to the evolution of cheese varieties were monasteries, manor houses, and feudal estates.

1.6   Monastic Cheesemaking

In addition to their roles in the spread of Christianity and the preservation and expansion of knowledge during the Dark Ages, the monasteries made considerable contributions to the advancement of agriculture in Europe, and to the development and improvement of food commodities, including wine, beer, and cheese. Many of our current well-known cheese varieties were developed in monasteries, e.g., Parmigiano Reggiano, Grana Padano, Wensleydale (Rievaulx Abbey, Yorkshire), Port du Salut or Saint Paulin (Monastery de Notre Dame du Port du Salut, Laval, France), Fromage de Tamie (Abbey of Tamie Lac d'Annecy, Geneva), Maroilles (Abbey Maroilles, Avesnes, France), and Trappist (Maria Stern Monastery, Banja Luka, Bosnia) (Rubiner 2016). The intermonastery movement of monks would have contributed to the spread of cheese varieties and to the development of new hybrid varieties.

1.7   Regionalization of Cheese Varieties

The role of the manor farms and estates and their breakup are described by Kindstedt (2012). The great feudal estates of the Middle Ages were self-contained communities. The conservation of surplus food produced in summer for use during winter was a major activity on such estates, and, undoubtedly, cheese represented one of the more important of these conserved products, along with cereals, dried and salted meats, dried fruits, dried and fermented vegetables, beer, and wine. Probably, cheese represented an item of trade when amounts surplus to local requirements were available. On these estates, individuals acquired special skills, which were passed on to succeeding generations. The feudal estates evolved into villages and some into larger communities. Because monasteries and feudal estates were essentially self-contained communities, it is apparent how several hundred distinct varieties of cheese evolved from essentially the same raw material, milk or rennet-coagulated curds, especially under conditions of limited communication. Traditionally, many cheese varieties were produced in quite limited geographical regions, especially in mountainous areas, where communities were isolated. The localized production of certain varieties is still apparent and indeed is preserved for those varieties with controlled or Protected Designation/s of Origin (PDO). Regionalization of certain cheese varieties is particularly marked in Spain, Portugal, and Italy, where the production of many varieties is restricted to a very limited region. Almost certainly, most cheese varieties evolved by accident because of a particular set of local circumstances. For instance, a peculiarity of the local milk supply, either with respect to chemical composition or microflora, an accident during storage of the cheese, e.g., growth of mold or other microorganisms. Presumably, those accidents that led to desirable changes in the quality of the cheese were incorporated into the manufacturing protocol. Each variety thus underwent a series of evolutionary changes and refinements.

The final chapter in the spread of cheese throughout the world resulted from the colonization of North and South America, Oceania, and Africa by European settlers who carried their cheesemaking skills with them. Cheese has become an item of major economic importance in the USA, Canada, Australia, and New Zealand, but the varieties produced are mainly of European origin, modified in some cases to meet local requirements.

For further information on the history of cheese, the reader is referred to Squire (1937), Cheke (1959), Davis (1965), Kosikowski (1977), Scott (1986), Kosikowski and Mistry (1997), Robinson and Wilbey (1998), Fox et al (2004, 2017) Kindstedt (2012), and Donnelly (2016). For references on Roman agriculture, see White (1970).

1.8   Standardization and Industrialization of Cheesemaking

Cheesemaking remained an art, rather than a science, until relatively recently. With the gradual acquisition of knowledge on the chemistry and microbiology of milk and cheese, it became possible to direct the changes involved in cheesemaking in a more controlled fashion. Although few new varieties have evolved as a result of this improved knowledge, the existing varieties have become better defined and their quality more consistent.

Considering the long history of cheesemaking, one might expect that the major cheese varieties would have been defined and standardized for a long time. However,

Table 1.1 First recorded date for some major cheese varieties (Scott 1986)

Gorgonzola	897	Cheddar	1500
Schabziger	1000	Parmesan	1579
Roquefort	1070	Gouda	1697
Maroilles	1174	Gloucester	1783
Schwangenkäse	1178	Stilton	1785
Grana	1200	Camembert	1791
Taleggio	1282	St Paulin	1816

although the names of many current varieties were introduced several hundred years ago (Table 1.1), these cheeses were not standardized; for example, the first attempt to standardize the well-known English varieties, Cheddar and Cheshire, was made by Joseph Harding in the mid-nineteenth century. Prior to that, Cheddar cheese was that produced in a particular area in England around the village of Cheddar, Somerset, and probably varied considerably depending on the manufacturer and other factors. Cheese manufacture was a farmstead enterprise until the mid-nineteenth century. The first cheese factory in Switzerland was established in 1815, the first in the USA was established near Rome, NY, in 1851 and that in Britain at Langford, Derbyshire, in 1870. Thus, there were thousands of cheese manufacturers and there must have been great variation within any one general type. This situation persists in a modified form today in Switzerland and Italy where there are a large number of small cheese factories, often grouped into *consortia* for the purposes of marketing and quality control. When one considers the very considerable interfactory, and indeed intrafactory, variations in quality and characteristics, which occur today in well-defined varieties, e.g., Cheddar, in spite of the very considerable scientific and technological advances, one can readily appreciate the variations that must have existed in earlier times.

Some major new varieties, e.g., Jarlsberg Maasdamer and Dubliner have been developed recently because of scientific research. Many other varieties have evolved very considerably, perhaps becoming new varieties, as a consequence of scientific research and the development of new technology. Notable examples are pizza cheese (a modified Mozzarella), (US) Queso Blanco, various cheeses produced by ultrafiltration, and various forms of quarg. There has been a marked resurgence of farmhouse cheesemaking in recent years; many of the cheeses being produced on farms are not standard varieties and some of these may evolve to become new varieties.

1.9 Milk for Cheese

A major cause of differences in the characteristics of cheese is the interspecies differences in the composition and physicochemical characteristics of the milk used. Bovine milk is by far the most important. Approximately, 85, 11, 2, and 2% of total milk is produced from cattle, buffalo, sheep, and goats, respectively. However, most sheep's and goats' milk is used for cheese and, therefore, are disproportionately important. Many famous cheese varieties are made from sheep's milk, e.g., Roquefort, Manchego, Feta, and the various Pecorino and Canestrato varieties. There are very significant interspecies differences in the composition of milk, which are reflected in the characteristics of the cheeses produced from them. Major inter-species differences of importance in cheesemaking are the concentration and types of caseins, concentration of fat and especially the fatty acid profile, concentration of salts, especially of calcium. There are also significant differences in milk composi-tion between breeds of cattle and these also influence cheese quality, as do varia-tions due to seasonal, lactational, and nutritional factors and of course the methods of milk production, storage, and collection.

The chemistry and physicochemical properties of milk and the principal inter-species differences are discussed in McSweeney and Fox (2013), McSweeney and O'Mahony (2016), and Fox et al. (2015). Cheese has been a popular research sub-ject for at least 100 years and a voluminous literature has accumulated, which has been reviewed in Mucchetti and Neviani (2006), Fox and McSweeney (2004), and McSweeney et al. (2017).

References

Barthelemy R, Sperat-Czar A (2001) Cheeses of the world. Hachette Pratique, Paris
Burkhalter G (1981) Catalogue of cheeses. Document 141. International Dairy Federation, Brussels
Cheke V (1959) The story of cheesemaking in Britain. Routledge & Kegan Paul, London
Davis JG (1965) Cheese, basic technology. Churchill Livingstone, London
Donnelly C (2016) The Oxford companion of cheese. Oxford University Press, London
Fox PF, McSweeney PLH (2004) Cheese: an overview. In: Fox PF, McSweeney PLX, Cogan TM et al (eds) Cheese: chemistry, physics and microbiology, vol 1, 3rd edn. Elsevier, Oxford, pp 1–5
Fox PF, McSweeney PLH, Cogan TM et al (eds) (2004) Cheese: chemistry, physics and microbiol-ogy, 3rd edn. Oxford, Elsevier
Fox PF, Uniacke-Lowe T, McSweeney PLH et al (eds) (2015) Dairy chemistry and biochemistry, 2nd edn. Springer, New York
Fox PF, Guinee TP, Cogan TM et al (eds) (2017) Fundamentals of cheese science, 2nd edn. Springer, New York
Harbutt J (2002) The world encyclopedia of cheese. Lorenz Books, London
Harbutt J (2009) World cheese book. DK, London
Hoffmann F (1761) A treatise on the virtues and uses of whey. L Davis and C Reymers, London, p 34
Kindstedt PS (2012) Cheese and culture. Chelsea Green Publishing, White River Junction
Kosikowski FV (1977) Cheese and fermented milk foods. Edwards Bros, Inc., Ann Arbor
Kosikowski FV, Mistry VV (1997) Cheese and fermented milk foods. F.V. Kosikowski LLC, Westport
MacAlistair A (1904) Food. In: Hastings J (ed) A dictionary of the bible, dealing with its language, literature and contents, vol 2. T & T Clark, Edinburgh, pp p27–p43
McSweeney PLH, Fox PF (eds) (2013) Advanced dairy chemistry, 4th edn. New York, Springer
McSweeney PLH, O'Mahony JA (eds) (2016) Advanced dairy chemistry, 4th edn. New York, Springer

McSweeney PLH, Fox PF, Cotter PD et al (eds) (2017) Cheese: chemistry, physics and microbiol-ogy, 4th edn. Oxford, Elsevier Academic Press

Mucchetti G, Neviani E (2006) Microbiologia e tecnologia lattiero-casearia. Qualità e sicurezza. Tecniche Nuove, Milan

Phelan JA, Renaud J, Fox PF (1993) Some non-European cheese varieties. In: Fox PF (ed) Cheese: chemistry, physics and microbiology, vol 1., 2nd. Chapman & Hall, London, pp 421–465

Robinson RK, Wilbey RA (1998) Cheesemaking practice, 3rd edn. Aspen Publishers, Gaithersburg

Rubiner M (2016) Monastic cheesemaking. In: Donnelly C (ed) The Oxford companion of cheese. Oxford University Press, Oxford, pp 486–487

Rubino R, Sardo P, Samuscal A (eds) (2005) Italian cheese—a guide to its discovery and apprecia-tion, 2nd edn. Slow Food, Bra

Sandine WE, Elliker PR (1970) Microbiologically induced flavours and fermented foods: flavour in fermented dairy products. J Agric Food Chem 18:557–562

Scott R (1986) Cheesemaking practice. Applied Science Publishers, London

Squire EH (1937) Cheddar Gorge: a book of English cheeses. Collins, London

Walter HE, Hargrove RC (1972) Cheeses of the world. Dover, New York

White KD (1970) Roman farming. Thomas and Hudson, London

Chapter 2
The History and Culture of Italian Cheeses in the Middle Ages

2.1 Introduction

In the 1980s, the French historian Michel Aymard (1983) highlighted how cheese was the poor relation in the extensive history of Medieval and Modern European food and farming. Aymard's observation may also readily be applied to the Italian context. According to Aymard, cheese was studied as a substitute or complementary food. It had a somewhat marginal role, not at all considered in itself worthy of attention.

Today, this understanding has changed profoundly. Italian historians of nutrition (e.g., Montanari 2008; Naso 1990, 2000; and Giagnacovo 2007, among others) have provided a far more refined understanding of the symbolic and cultural characteristics of cheese, exploring also the importance of its trade and consumption in the Middle and Modern Ages. The revisions and clarifications promoted by the above mentioned scholars' historical analysis have allowed identification of specific cheese protocols down to regional level.

For those from other disciplines unfamiliar with the historical aspects, it would be useful firstly to summarize the state of the art for the period prior to the Middle Ages. This includes description of the extraordinary richness and variety in localized Medieval Italian cheesemaking, with comparative reference made with the wider Mediterranean and transalpine areas.

2.2 The History of Cheese in Italy to the Eleventh Century

Although cheese is constantly present in the texts from the ancient Western world (see also Chap. 1), it is clear that it had negative connotations. When Ulysses entered the cave of Polyphemus, with his companions, he found buckets of milk and cheeses laid out on grates for drying. The giant cyclops was a hideous and uncivilized

© Springer International Publishing AG, part of Springer Nature 2018
M. Gobbetti et al., *The cheeses of Italy: Science and Technology*,
https://doi.org/10.1007/978-3-319-89854-4_2

shepherd, with his dairy practice as evidence. As soon as Polyphemus returned home, he started milking his animals and then preparing the curd. Such negative interpretations are also found in the writings of Herodotus, the father of history. Herodotus describes of the Scythians with disdain as being milkers of mares (Montanari 2011). In the ancient Greek mentality, cheese represented something reprehensible. People living a primordial life, and having a passive and subordinate attitude with respect to nature, had the custom of drinking milk. On the contrary, civilized people produced bread, wine, and oil, demonstrating their capacity to create and develop these technological processes.

The Latin culture began to move beyond these ancestral ideas. Tibullus and Virgil idealized the primitiveness and innocence of previous historical periods of civilization. A different understanding developed, overlooking apparent contradictions developed. An example is Polyphemus, as the oldest reference to cheesemakers, yet cheese is the fruit of human ingenuity. When the Romans encountered the nomadic populations of herdsmen from Africa and central-northern Europe, they noted the herdsmen's total ignorance of agriculture. For the Romans, eating such unrefined foods as curdled milk and meat represented an index of barbarism and backwardness. Evidence of this is found in Sallustius, who judged the Numidians to be barbarians; with Caesar and Tacitus stating similar about Gauls and Germans. Pliny the Elder (first century AD) expressed astonishment that populations living *ab immemorabili* on milk did not know cheese. For Pliny, cheese was a product created by human genius, along with other foods; this represented an important step forward.

Caseus ("cheese") was widespread in Roman society. Nevertheless, only the poor or soldiers consumed cheese. Roman nobility generally avoided *caseus*, except as an ingredient for preparing particular dishes. Agricultural and economic treatises from the republican and imperial era (Cato the Younger, Varro, Columella, and later Palladius) report much information on cheesemaking, and related aspects such as quality and trade, strongly indicating its importance. At that time, most cheeses were made from cow's milk (the most nutritious but least digestible milk), sheep's milk, and goat's milk. Usually, soft and fresh cheeses (*casei molles ac recentes*) were destined for the nearby urban markets and for immediate consumption. These were distinguished from hard and aged cheeses (*arid et veteres*), which were expensive but also transportable over very long distances. Cheese mainly came to Rome from the Apennine regions and southern of Italy, but also from Sardinia, Provençal, the Alps, Dalmatia, and Asia Minor. Among these, the cheese from "Luni" deserves a special mention. The identification of this cheese was thanks to a large moon stamped on the cheese: *caseus Lunensis Etruscae signatus imagine lunae* (cheese from Luni, stamped with the Etruscan moon) (Amiotti 2011).

Continuing this line of enquiry, there are further descriptions of the use of rennet from various animal and vegetable sources, and referring to the gastronomic and therapeutic uses of various cheeses. Concerning the cheese trade, it is worth briefly mentioning the first Virgilian eclogue. The shepherd Titirus complains about the low price he could extract from the *ingrata urbs* at the market, for his considerable amount of cheese.

2.3 The Middle Ages

2.3.1 The Early Middle Ages

The early European Middle Ages is a period of interaction between Roman civilization and the barbarian world, through the slow and gradual process of Christianization. About the history of cheese, little is known, since there is little extant documentation surviving from before the eleventh century; written testimonies are rare and archeological sources almost none.

Undoubtedly, the interrelationship between people and cattle was central to barbarian societies and for the Germanic peoples, in particular. With the term Germans, Romans defined a myriad of tribal groups, who remained for the longest period without a self-identified ethnic identity. Sturleson's *Edda* talks of cosmogony, and how the cow *Audhhumla*, meaning a prosperous hornless cow, played a decisive role within the primordial event (Bellini 2011). Other traces exist, for example, Irish druids described some cheeses and the use of curdled milk as a food, although beyond this little is known for certain. The edict of Rothari (643), from the politically and socially dominant Longobards in central-northern Italy, placed great emphasis on the importance to them of sheep, goat, cattle, and pig breeding. Although precise confirmation is still lacking, Longboards are generally seen as having imported buffalos to Italy.

Ecclesiastical sources provide information about cheese manufacture and consumption in late antiquity and the early Middle Ages. The main literature concerns penitential books, which established the canonical penalties for clerics and laity, theological texts, monastic, and, later, Carolingian literary sources. A few cursory references are outlined as follows. The punitive diet for priest guilty of sexual sins was the consumption of milk, instead of wine, and the replacement of meat with the Breton *formellus*. When the prescriptions become more severe, the consumption of these two dairy products was also prohibited (Bellini 2011). From theological sources, the *Passion of St Perpetua* (third century AD) narrates a suggestive vision. Having escaped prison with her companions, the future Saint Perpetua met a white-haired man, dressed as a shepherd and milking sheep, who offered her *de caseo quod mulgebat* to eat. The saint received this dairy product (probably Ricotta or curdled milk) with hands clenched, and the attendants said *amen* (Bellini 2011). The evangelical and Eucharistic resonances of the "good shepherd" are obvious. The complexity of the cultural processes between the fourth and fifth centuries (at that time St. Augustine wrote the *liber de heresibus*) is clearly shown through the heretical definition of the ritual of *artotyriti*, who, as their Greek name suggests, used bread and cheese to celebrate their mysteries (Bellini 2011). This liturgical practice had its foundation also in the New Testament. Even if the Gospels never refer to cheese, some of the famous passages of the first letter to Corinthians and the letter to Hebrews used milk to describe some metaphors. Milk was the spiritual food given to believers, having weak and infantile minds. On the contrary, solid foods were for aged and mature men. The medieval explanation of the Scripture did not

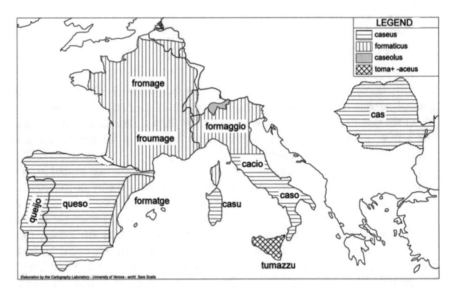

Fig. 2.1 Names for cheese in the Romance languages. Continuations of Latin *caseus* prevail in peripheral areas less exposed to innovation (Portugal, Romania, southern Italy), as well as in non-Romance languages (German, English). In the area corresponding to the Carolingian Empire (from Catalonia to the Rhineland), the continuations of *formaticus*, from *forma*, is found. The exception is the name for cheese in Sicily, where the "*tumazzu*" form prevails, derived from the term "*toma*," which is of uncertain etymology but quite widespread in local dialects, in Northern of Italy also. (From G.L. Beccaria, Tra le pieghe delle parole. Lingua storia cultura, Einaudi, Turin 2008, p. 34, derived in turn from research by the German linguist Gerhard Rohlfs)

escape this comparison. Indeed, it established a parallelism between whey/cheese/butter and literal/allegorical/moral interpretation of the scriptures. In southern Italy of Byzantine-Greek tradition, usually prayers accompanied the cheese blessing. This tradition remained in use until the Latinization of the Norman period. Far more important were the liturgical and iconographic celebrations of milk: milk and honey are deposited on the lips of baptized children and there is the image of breast-feeding Madonna.

Monastic and ecclesiastical sources provide more precise and a greater amount of data. The first and most important event confirmed the distinction between cheeses, based on protocols for manufacture and variety (Beccaria 2007). The linguistic map of romance languages in European (Fig. 2.1) attributed the *formaticum/fromaticum* varieties to the Carolingian territory (France, northern of Italy, Alps, and Catalonia). Varieties of *caseus* were from the Italian peninsula, Sardinia, and lateral areas (Romania, Portugal, and Spain). *Tuma/tumazzu* (from *toma*) varieties were from Sicily, with *formaticum* representing their alternative. Studying the inventories of goods and the administrative documents by monasteries of the Middle Age, it emerges that *caseus* indicated fresh cheese. *Formaticum* (or *caseus de forma*) presupposed the use of ripening boxes, which allowed whey syneresis and gave shape to the cheese. Unsurprisingly, the commercial estimation of *caseus* was in pounds, while that of *formaticum* was in pieces of a standardized shape. Correlating age and flavor, the *formatici boni* were the most mature ones.

Important information also comes from the ninth century collections of the norms of daily life (*consuetudines*) of individual monasteries or congregations. Usually, the cheese was stored in monastery cellars, where some theft also occurred. During Lenten fasts, monks were prohibited from consuming cheese. In the Byzantine era, the week of cheese or dairy products came into use as a sort of pre-Lent training (Parenti 2011). Nevertheless, cheese was always a simple food, mainly for shepherds and peasants. It was also one of the main components of monastic diets, as an alternative to eggs, especially when the consumption of cooked foods was prohibited. During the eleventh to twelfth centuries, the most important findings concern the special attention given to cheese varieties. As in Roman times, the classification of cheese was through adjectives describing its origin (e.g., Flemish cheese, Alpine cheese). The story told by Notkero Balbulo in the *Gesta Karoli Magni* is worthy of note. On a day of fasting, a Benedictine abbot offered the emperor Charlemagne a fat white sheep's cheese, covered in whitish mold. The abbot warned his illustrious dinner companion that removing the moldy rind would take away the best part of the cheese. Charlemagne so greatly appreciated the cheese that he requested an annual supply; however, the expert abbot cautioned that moldy cheeses would not always have the same taste (Nada Patrone 1996).

2.3.2   The Late Middle Ages (Twelfth to Fifteenth Centuries)

Two closely interconnected and equally decisive structural elements make another and more analytical description possible of the history of cheese in Medieval Italy.

The first element concerns demographic developments. Even in the economic crises of late antiquity, Italy and its provinces continued to be the most densely populated and urbanized area in Western Europe, deriving from its Roman heritage. From the tenth century, although possibly even earlier, the trend in urbanization grew uninterruptedly for 400 years, until the beginning of the fourteenth century. Around the fourteenth century, at the height of demographic developments, a systemic crisis occurred. Several factors have been determined for this crisis, not least intense agrarian development and the marked reduction in forests and pastures. Exemplary of the intense urbanization, Central and Northern Italy had at least 30 cities with more than 15,000 inhabitants. This is without even considering the big metropolises such as Milan, Venice, Florence, Bologna, and Genoa. The demand for food, in a structurally weak agrarian system, increased. The weakness of the system was due principally to modest agricultural yields, based on the extent of the cultivation rather than on crop intensity. Primarily, each city produced foods from its own territory (often coinciding with the diocese) and, consequently, developed its own livestock. Nevertheless, large metropolises, without an agrarian hinterland (e.g., Venice), were powerful engines for market development. Research over the recent decades demonstrates how the expansion of large-scale European trade, together with improvements in navigation, at sea and inland, all encouraged an intense commercial network for food, especially in the center and north of Italy. This trade in food also included ripened cheeses. The major repercussion was on the lowering of transport costs, especially over long distances, within the Mediterranean area. This is considering that in 1191, only one Venetian ship had the capacity to carry 22,000 cheeses.

The second element, which influenced the history of Italian cheese in the late Middle Ages, is the extensive presence of written documentation. In Italian cities, where the municipality became the main form of government, notaries produced private and public documents that also referred to cheese varieties. This added to the more limited ecclesiastical sources about cheese varieties, which had been known previously. Many reports on license contracts for pastures came out. The payment of these licenses was by money or through some percentage of the product. Legal sources were also important. In particular, witness testimony, statutes of cheese vendor corporations (*casolini* and *formaierii*, which conferred an elevated rank within the city's hierarchy) and city statutes, all reported useful information. City statutes strictly prohibited the use of ripened cheeses for playing the game *ruzzola*, which was a very common pastime consisting of rolling cylindrical cheeses in the street. Later, sources come from private documents of individual families (mainly bourgeois and merchants), records from hospitals, merchant manuals (*pratiche di mercatura*, dealing with commodity and measure units), and city duties. In particular, the latter made a classification of cheeses, thus identifying fundamental varieties.

At that time, the ecclesiastical sources maintained their importance. Thanks to pastoral visits or sermons, they testified to the social elevation of cheesemakers and livestock breeders. Canonists from the eleventh and twelfth centuries (Burcardo and Ivo of Chartres) still depicted cheesemakers and livestock breeders as living on the border of society, even on the border of the rural society. One of the most famous definitions was as follows: "cow and pig farmers and shepherds live on the borders of society, like beasts, and mainly populate the forests"; nevertheless, "Christ also gave redemption to farmers and shepherds with his blood." By the late Middle Ages, it was not rare to find cheesemakers as leading members of society in Italian cities.

Lastly, the scientific treatises represented other significant sources of information. The *Summa lacticiniorum*, by the Piedmontese doctor Pantaleone da Confienza, was a milestone, not only in Italy but for Europe (France, Flanders, and even England). In the second half of the fifteenth century, this text provided an overview of the state of the art of cheesemaking. The final section will describe in depth the *Summa lacticiniorum*.

2.3.2.1   Farming and Cheesemaking in Northern Italy

The process of anthropization induced by the demographic and economic developments of the eleventh to fourteenth centuries had important consequences on farming systems on the Italian peninsula. This included the Po Valley, the Alpine and pre-Alpine mountains, the Italian Apennine, the coastal plains of the center and south of Italy, and the Mediterranean islands. The most important novelty in the Po Valley was the development of cow farming, which accompanied that of sheep, already present because of the remarkable demand for wool. It was a very long process, which became appreciable only in the late Middle Ages (fourteenth and fifteenth centuries), giving impetus to the overall environmental and agrarian meta-morphosis. The soil surface without agrarian cultivations diminished markedly. In some areas of the Valley (especially in Lombardy, but also in Veneto), the construc-tion of irrigation canals created large areas of grassy meadows, which would have allowed the increase of forage production and the consequent practice of farming. The southern slopes of the Alps and the pre-Alps were fully involved in this devel-opment, thanks to the transhumance from valley to mountain, even over distances of many kilometers. Permanent and seasonal farms expanded upwards to an altitude of around 1800–2000 m. Monasteries (especially Cistercians in Piedmont and Lombardy), together with rural and mountain communities, were owners of valley pastures or used valley pastures under usufruct arrangements. These groups became the most important actors in the extraordinary development of the pre-Alpine and Alpine pastures (Fig. 2.2). Undoubtedly, the economic and political context of the city had a primary role on this evolution.

At that time, Italian geography included very few homogeneous mountainous territories. The agrarian expansion over 800 m in altitude was found only in the val-leys of Aosta, Trentino, Tyrol, Cadore, and Carnia. Because of these geographical conditions, it is difficult to apply the same model that the French historian, Arbos, usefully proposed for classifying the cheeses from the French and Swiss Alps. Arbos made a comparison between the *grande montagne* (with public pastures) and the *petite montagne* (with private pastures). The first was the *montagne à gruyère*, which needed large amounts of milk daily (and therefore large numbers of cows) for making large, heavy cheeses. The second one was the *montagne à tommes*, where the owners produced small, dried cheeses (2–10 kg) using small amounts of milk. This scheme, also used for Emmental cheese and, more recently, for Fontina cheese from the valley of Aosta (Comba 2011), is very useful for classifying antique cheeses. Another appropriate example referred to the Savoy jurisdiction of Quart (*castellanìa*) (1377–78). Here, the distinction was between fat-containing cheese and *seràs/sérac* (nonfat cheese similar to Ricotta). The manufacture of the latter was from cooked whey, after the preparation of fat-containing cheese using long paral-lelepiped forms called *rucze* (Pession 2015; Naso 1999).

In Northern Italy, reports on farming/transhumance/pasture intersected the eco-nomic system of the city, and the consequent cheese denominations relied mostly on individual cities. Several factors linked to territorial characteristics interfered with cheese production and trade. In the Padania inland, for instance, the most important of these factors was the distance between Alpine valleys and cities. The distance from Brescia to reach an altitude of 1000 m was around 90–100 km, while the dis-tance from Verona or Vicenza to the mountains was only 20 km. Undoubtedly, the shorter distance favored the daily supply of fresh cheeses to urban markets. The legal customs of mountain populations was another influencing factor. Walser

Fig. 2.2 Dairy farming in the mountain pasture. The four separate operations, which follow each other consecutively, are depicted together: milking the cows, transporting the milk, manufacturing the cheese, and manufacturing the butter. (Month of June, from the fresco of the Cycle of the Months in the Torre dell'Aquila at Castello del Buonconsiglio in Trento; end of the fourteenth century) (Photo A. Bednorz, 2009—© Castello del Buonconsiglio, Trento)

immigrants from the western Alps or German-speaking settlers from Venetian pre-Alps were owners of Alpine pastures. This contrasted with forms of collective possessions by communities from Italian villages and valleys. This conflict of interest often determined a decrease in the availability of farms and dairy holdings. In the Alpine territory, cattle from a single village were herded communally, with only a single *casera* (dairy production unit) for shared use.

In such an extremely fragmented context, caution is necessary in ascertaining the veracity of information (cheese name, quality, and trade) coming from single cities. Different definitions, sometimes transient or sometimes destined to last, made sense within the local system of denomination, but not in general terms. Specific examples support this methodological consideration. In Treviso (thirteenth to fourteenth centuries), the classification of cheese relied on distinctive pairs of terms: fresh or ripened; cow's or sheep's; *nostrano or forese* ("domestic" or "foreign"); salty or sweet. Similar adjectives distinguished the cheeses from some Friuli cities: at Gemona, *latinus* or *teutonicus*; at Cividale del Friuli, *vetus salsus* or *dulcis novus*, and again *teutonicus*. The cheeses sold in Piacenza's market had the general definition of *caseus nostranus*, which, however, referred to specific organoleptic characteristics. Merchants from this Emilian city went to other cities to buy homemade cheese, *causa emendi caseum nostranum* (Greci 2011). It is also possible that another cheese, *caseum nostranum seu grassum* (Naso 1996), originated in Piacenza (twelfth to thirteenth centuries), although its identification is very uncertain, being limited to the adjective *grassus* (Figs. 2.3, 2.4, and 2.5).

Not only those manufactured in the territory but also cheeses of very distant origins circulated in Northern Italy at that time. This is the case of *sardesco* or *de Creto* cheeses from Sardinia and Crete, respectively. Nevertheless, the qualitative attributes seem to be rather different from those of the same cheeses made in the above islands. There are reports that Sardinian cheeses (*sardinalis*) were present also in Rome during fourteenth and fifteenth centuries and in Bergamo (Bonazza 2011). In Tuscany, where Sardinian cheeses circulated much more abundantly, there were at least five or six varieties.

Further examples of cheese classification relied on cities, which were geographically contiguous in Brescia territory (Archetti 2011). *Caseum de consilio* was the cheese given as remuneration for public council functions of the rural town hall. *Caseum sanctuarie* was the cheese made with the milk provided only on the feast day of Saint John Baptist. The identification of *caseum scumarie*, a semi-finished creamy Ricotta obtained from the foam of the boiled whey in the *caldera*, was rather difficult (Archetti 2011). This denomination was not replicated elsewhere. The *mascherpa* (*mascharpa*), a salted cow's milk Ricotta consumed fresh, ripened or smoked (Archetti 2011), was found elsewhere in Lombardy (Bergamo). The main characteristics of *matellus/mathellus* cheese remain unknown, with its presence documented only between the territories of Brescia and Bergamo. In the neighboring area of Verona (thirteenth century), at least five denominations indicated cheeses manufactured in the pastures of the Lessini Mountains. These denominations were completely unknown elsewhere, including all the other mountains of the Venetian region. Documentation also exists concerning the cheese denominations

Fig. 2.3 Milking sheep. This image and those that follow (Figs. 2.3, 2.4, 2.5, and 2.6) are taken from the *Tacuinum sanitatis*, ms. 2644 of the Österreichische Nationalbibliothek in Vienna (end of the fourteenth century; formerly owned by George of Liechtenstein, bishop of Trento and commissioner of the fresco of the Cycle of the Months, of which a detail is reproduced in Fig. 2.2). The *Tacuina sanitatis*, of which there are numerous copies (preserved today in various European libraries: in Rome, Paris, Liège, etc.), constitute the Latin translation (with illustrations) of an Arabic text dating back to the eleventh century, written by the (Christian) doctor and writer Abu al-Hasan al-Mujtar ibn Butlan, originally titled *Taqwim al sihha* ("Health Almanac"). A large part of this work is dedicated to foods. For each food, brief medical notes are provided: the complexio or nature of the food that characterizes its best state (*electio, melius ex eo*), the advantage that the body derives from it (*iuvamentum*), the discomfort (*nocumentum*) that consumption may cause a certain type of person (young or old, male or female, etc.), and the remedy to be used (*remotio nocumenti*). These recommendations are given based on the calendar (spring, summer, autumn, winter), in accordance with the theory of humors typical of medieval medical thought (sanguine temperament in spring, choleric in summer, melancholic in autumn, and phlegmatic in winter)

Fig. 2.4 Fresh cheese (*caseus recens*). As in other *Tacuina*, the vignette depicting fresh cheese in the Vienna *Tacuinum* represents the moment of preparation. Under the wooden porch of the hut, the peasant girl works the curd, pressing with her hands inside a circular wooden shape. The excess whey flows from the table into a large basin, where the dog eats it. Moreover, the caption explains that the benefit of this food is that it "softens and fattens the body"; however, it causes occlusion and constipation, which can be remedied by consuming walnuts and almonds

Fig. 2.5 Aged cheese (*caseus vetus*). In all the *Tacuina*, the image of a town shop is reserved for aged cheese: the shelves with large shapes (all the same), the scale and the cat. This suggests that this type of cheese, more than any other, was the subject of medium and long-range trading. The caption explains, among other things, that aged cheese is very nutritious, and it is better for young people who do strenuous jobs than for elderly people. It can damage the kidneys, cause occlusion, and produce kidney stones; it must be consumed between two meals. However, eating horseradish the day after consuming it remedies these inconveniences

Fig. 2.6 Ricotta (*recocta*). For this product, the Vienna *Tacuinum* simultaneously depicts the moment when the product is produced (the milk is on the fire in a large cauldron) and tasted: probably both *ricotta salata* (by the man on his feet) and fresh ricotta (by the young man, who takes it with a spoon). The equipment (strainers, ladles) is depicted carefully. From the dietary point of view, it is noted that the product is difficult to digest (to be remedied by consuming butter and honey). It is nevertheless nutritious and suitable for young people who eat it at the start of summer and in mountainous regions

ad pueros (for children) and *oculos* (cheese with holes). The term referring to holes indicated very poor quality, as it is a technical term still in use. Other examples were *Alferinus* cheese (by the name of a mountain village, Alferia, currently Cerro Veronese), *maçaegus* cheese (because of its manufacture during the month of May), *braçatica* cheese (recalling the concept of arms), and the *povina asinaria* (Ricotta of low quality, with the name expressing disdain for donkey) (Vigolo 1990).

At end of the fourteenth century, the extraordinary archives of Francesco Datini, a merchant from Prato, supply a great deal of information. These archives reported the main consumption and trade of a generic cheese (fresh or salted) in several neighboring cities (Pisa, Florence, Genoa, and Avignon). The *parmigiano* or buffalo cheese were in second place for consumption in Italian cities. In Avignon, *tome* cheeses were famous (Giagnacovo 2007).

Without other definite clues, it would be prudent not to identify current cheese varieties with names from medieval sources, if these names only refer to an indication of places of manufacture. The current commercial systems (e.g., websites) are always looking for tradition and naturalness, claiming the genuineness of former times. However, in several cases, they refer to the invention of tradition. For instance, it was sufficient that a document from 1277 stated that the marquis of Saluzzo (Piedmont) received a generic cheese as payment for his commitment to state further that this was Castelmagno cheese. Likewise unfounded is that the manufacture of Asiago cheese began around the eleventh century, since it was an upland zone that was largely uninhabited at that time. The almost certain identification of a medieval or modern age ancestor if a current cheese needs to satisfy three criteria: to get almost the same information from different sources; such sources must be contemporaneous; and at least one organoleptic feature must be easily recognizable and consolidated.

Undoubtedly, one of the few varieties to have met the above criteria since the late Middle Ages is a cheese, ripened and grated, similar to current Parmigiano Reggiano or Grana Padano. At that time, Piacentino and Lodigiano cheeses with almost identical characteristics were equally famous (Montanari 2005). In the second half of the thirteenth century, Salimbene de Adam (a Franciscan chronicler from Parma) described the use of a grated cheese as a pasta flavoring ingredient in this area. The mention of *parmigiano* in the Decameron is very famous. One of the major attractions in the country of Bengodi is the "whole mountain of grated *parmigiano* cheese, on which people stayed making only macaroni and ravioli, cooked in capon broth" (Montanari 2011). At the same time (1351), the statute of Bologna referred to *formagias casei parmensis*. By the end of the fourteenth century, the wide circulation of this cheese is attested to. In 1389, Pisan merchants were loading big cheeses (around 13 kg, and later 20 kg, which were more impressive from an aesthetic point of view) on their ships, for transport in France, Spain, and even North Africa. All this indicates that at least in this area there was a considerable production of cow's milk, derived from farms without long-distance transhumance, but with large numbers of cattle. Protocols for cheesemaking were already present in fourteenth century documentation, including the Piedmont territory (Naso 1996).

2.3.2.2 Farming and Cheesemaking in Central Italy and the Apennines

The geographical formation of Central Italy, and in particular the Apennine ridge, imposed the prevalence of cheeses made with sheep's milk and goat's milk. This was completely different from the cheesemaking conditions of Northern Italy. In the center of Italy, cows were mainly working animals, although their farming was not completely absent in the zones of Tuscany and Lazio. Clearly, this has been the situation since ancient history. It is not a coincidence that Pliny the Elder, in his review on cheeses from the Italian peninsula, documented five Apennine productions, with only two for the Alps. In addition to the "Luni" cheese mentioned previously, there were cheeses from Ceva (Ligurian Apennines), Sarsina (between Umbria and Romagna), and two from two cities of Abruzzo or Campania (the Vestini territory, the *campus Caedicius*).

The Middle Ages were a period of incisive novelty. In particular, the development of trade along the seacoasts favored the imports. The most important imports were from Arabian countries of the Mediterranean coast, Sardinia towards Pisa, and from the Balkans towards the Adriatic cities. The production of cheese in Croatia and Serbia was enormous. According to Michele Savonarola (fifteenth century), a then-famous doctor hostile to cheese, and its abundant consumption, influenced the character of the Croatia and Serbia inhabitants. Indeed, it was believed that those who consumed salted cheeses from childhood became choleric, bothersome and were with bad blood (Camporesi 1990).

The aforementioned documentation from the late fourteenth century, in the Datini archives in Prato, mention almost a dozen specific cheese varieties. Five of these varieties were certainly imported from Northern Italy or from the Mediterranean islands: *sardesco*, *calorese* (probably *gallurese*, from the Sardinian Gallura), *lombardo*, *di Locarno* (Canton Ticino, Switzerland), and *parmigiano*. Other cheese varieties were regional, such as Pisan cheeses *marzolino* and *raveggiolo*. Like other sources, the archives of a merchant from Prato also reported the cheese from Lucardo (a small center in territory of Siena) (Giagnacovo 2007), which was a rare case of naming a variety after the village from which it originated. Nevertheless, most of the denominations reported in these archives came from commercial letters and were too generic and difficult to identify geographically. Examples of these generic denominations were *pettinato*, *cavallino* (purchased in Palermo and possibly corresponding to a Sicilian cheese), buffalo fat cheese, fresh, dry, and salty cheese, Ricotta, and *cacio di forma*.

To these generic names, it would be easy to add other varieties, which did not have a character indicative of Central Italy. One example is the *giuncata* or *felciata* (from ferns) cheese, produced in the Apennines (Marche region), but also produced and spread widely throughout Italy. The manufacture of this cheese was with fresh curdled milk, giving a creamy consistency. The name *giuncata* or *felciata* derived from the box for molding (*fascera di giunchi*) or from the material (fern leaves) used for milk coagulation. According to historians, the name *giuncata* coincided with the Piedmont *seracium*, also called *mascarpa* in Lombardy (Nada Patrone 1996), all made from milk whey. Nevertheless, the *giuncata* cheeses were not always distinguishable, based on organoleptic features, from other similar varieties such as the *capolatte* (produced from long boiled milk), or the *lattimele*. In any case, the *giuncata* was a perishable cheese for immediate consumption, so that transport to the urban market had to occur daily. As recorded, the *giuncata* had a proverbial velvety softness and an extremely delicate flavor. In *Don Giovanni* by Mozart-da Ponte (1787), *giuncata* is the term used to describe the freshness of farmer Zerlina's complexion, as stated by her seducer (*parmi toccar giuncata e fiutar rose*).

Among the above varieties, and excluding the other scarcely definable cheeses (e.g., *nostrano*, *aretino*, and *pisano*), *marzolino* and *raveggiolo* were the most important sheep's cheeses manufactured in central Italy. The preferred consumption of *marzolino*, which derived its name from the season of production (*primavera*, spring), was at the beginning or at the end of the meal, together with fruit. The sale was in pairs of various sizes, tied together, and storage was in oiled cloths to keep its softness. The extraordinary Tuscan correspondence (fourteenth and fifteenth centuries), especially from women (Margherita Datini and Angela Macinghi Strozzi), showed the consumer's need of having some shrewdness before purchasing cheese. "It was neither beautiful nor good," wrote Margherita Datini with disappointment about a batch of cheese. Therefore, assessment of cheese before purchase presupposed a direct knowledge of the site of production, and of the specific rural cheesemaker, or to be acquainted with those who know where good *marzolino* cheese was sold (Muzzarelli 2015). A specific description of *marzolino* cheese emerged mostly in the sixteenth and seventeenth centuries, when the Siennese and Florentine literati constructed the Tuscan "tradition," presenting *marzolino* as the cheese of the Grand Duchy of Tuscany *par excellence*.

Raveggiolo was a *giuncata*-type cheese made from goat's milk, was fresh and soft, with a squashed shape (Giagnacovo 2007). In the sixteenth century, the famous chef Cristoforo Messisbugo listed the cheeses that "cannot fail ... for the coming of every great prince" to court (in this case, the court of Ferrara), including *raveggiolo*, together with *marzolino*, *provatura*, and other fresh and ripened cheeses.

In the context dominated by Apennine cheeses, the exceptions were the cheeses from lowland areas on the Tyrrhenian side. Here, the political power managed the pastures and created the so-called *Dogane* (customs body). The *Dogane* provided some facilities for shepherds, including for the purchase of salt. Examples of pastures in the Tyrrhenian area were the *Maremma senese*, partially covered by swamps and suitable for sheep and cow farming; in the Papal States, the *Campagna romana* and the lands of the Patrimony. As far as the *Maremma senese* is concerned, it is significant that already in the early thirteenth century, the city's documents considered cow's, buffalo's, and sheep's cheeses (*cascio*) at the same level. The *Campagna romana* had particularly new types of residential and productive emplacements, also fortified with towers, which are still traceable in the current toponymy (e.g., Tor di Valle, Torrimpietra). These settlements were the so-called *casali*, controlled by Roman barons. However, a new and dynamic economic class of farmers and ecclesiastical corporations constructed and managed the *casali* (Maire Vigueur 1974; Carocci and Venditelli 2004). At the beginning of the fifteenth century, the Pope established the *Dogana del bestiame* (livestock customs body), thus centralizing the pasture management in the *Provincia del Patrimonio* (north of Rome, close to Viterbo and Tuscany), where hundreds of thousands of sheep and cows grazed during winter (Maire Vigueur 1981). The cheese sector was remarkably important in the context of the economic recovery of Rome in the late fourteenth and especially fifteenth centuries, when the Pope returned to Italy. In the 1452–53, the export of cheese alone accounted for about two-thirds of the total volume of urban exports in Rome. Excluding imported cheeses (Sardinian *marzolino* cheese), the most common cheese varieties were the buffalo's milk *caseus bufalinus*, *bubalinus*, or the *provatura* fresh cheeses (Lanconelli 2011; Lombardo 1983), and the not-so-well-identified *cacio di forma*.

2.3.2.3 Farming and Cheesemaking in Southern Italy and the Islands

Part of the cheese manufactured in continental Southern Italy and in the Mediterranean islands was destined for domestic markets, including the major cities of Naples and Palermo. Nevertheless, in the Middle Ages the largest part of the cheese production was for export. Through small and big ports spread along the Adriatic and Tyrrhenian coasts, cheese went by sea to the key emporiums of Central and Northern Italy (Venice, Genoa, and Pisa). From here, the cheeses were redistributed to other cities. The volume of traffic was enormous. In Genoa, there was a rare but significant example of a shop for cheese wholesaling and retailing (*ripa Formaieriorum*). In this place, only in the week from 9th to 15th May, 1269), 28 tonnes of Sardinian cheese were sold to merchants from Florence and Lucca. Nevertheless, the Italian market was not self-sufficient, part instead of a broader Mediterranean market (Basso 2011). Venice was the port of arrival for Apulian and Sicilian cheeses, but also for cheeses from *Barberia* (Egypt), Albania, and, especially, Crete (Basso 2011). The production of sheep's cheeses from Crete was also considerable, reaching the Byzantine and Eastern markets. Progressively, the cheeses from Crete lost their appeal to higher-class Western consumers, whose refined tastes no longer appreciated these varieties, due to their saltiness. Cretan sheep's cheese was mentioned in the manuscripts of Venetian merchants, who described that this cheese lost ca. 7% of its weight during navigation. Ripened cheeses were also essential components of the mariner's diet.

In the middle centuries of the Middle Ages (tenth to twelfth), the most important information on farming and cheese manufacture from the continental *Mezzogiorno* (or south) of Italy came from the archives of important monasteries (Montecassino, Cava dei Tirreni, Montevergine, S. Sofia of Benevento, and S. Maria delle Tremiti). Sometimes, these literary sources indicate diversification from producing the predominant sheep's cheeses to raising cattle, mainly in Apulia (Capitanata, in the province of Foggia) and Campania (mainly in the lowland formed by the Sele river) regions. From the thirteenth century, the court of Naples became an important center for cheese consumption and production. Nevertheless, during the Angevin period, French gastronomical chauvinism recommended the use of non-Italian cheeses. One example dealt with the *de Bria* cheese, largely used to prepare a

Gallican soup. In the mid-twelfth century, market expansion in this territory was very great. Annually, the expectations by royal administrators were for approx. 800 kg of cheese and approx. 160 kg of Ricotta for each flock of 100 sheep (Di Muro 2011).

In fourteenth and fifteenth centuries, initiatives by the king led to significant developments. Already at the time of the Norman-Swabian kings, continuing with Frederick II, the Apulian lowlands (the Capitanata and Tavoliere delle Puglie) were being repopulated and developed into large-scale farms known as *masserie*. This farming system played an important role, especially after the effective abandonment and near total depopulation of the High Middle Ages, which began in the fourth and fifth centuries. In the fourteenth century, various actors (ecclesiastical institutions and the same Angevin monarchy) and other factors (including the demographic crisis of the second half of this century) led to a profound transformation, also at the environmental level, which favored the development of the transhumance of sheep. This led to the organization of a complex, efficient, and profitable public body, directly dependent on the Aragon monarchy: the *Dohana mene pecudum* (sheep transhumance customs authority). The *Dohana mene pecudum* had its location first in Lucera and then in Foggia, and managed millions of sheep per year. Since that time, and for the subsequent four centuries, the economic system of the Apulian lowland became *une usine à laine, à fromage, à viande* (a farm for wool, cheese, and meat) (Martin 2007; Violante 2009). The accounting records for this period (beginning in 1452) indicate remarkable productions of *caciocavallo*, ricotta, and simple *cacio*.

The major share of cheese produced in the Adriatic regions found its way to Venice (together with the olive oil and wheat from Apulia). The circulation of cheeses in the Tyrrhenian Sea, and especially from Sardinia and Corsica, was well documented. Since these two islands were underpopulated, a colonial economic framework was established, with two main destination markets. The first was Catalonia and the Iberian Peninsula. At the end of the fifteenth century, 90% of the fresh or salty cheese traded in Valencia came from Sardinia or Majorca (Basso 2011). The Tyrrhenian cities (Genoa and Pisa, but also Naples, Salerno, Trapani, Majorca, Tunis, and even the Calabria region) (Galoppini 2006) represented the other destination. The importation of a generic Sardinian or *sardesco* cheese was documented in these cities. Occasionally, one or two other adjectives were used with this cheese: *sardesco* white, and *caseus salsus,* or *freschus Sardinie*. In other cases, the same cheese was identified with the subregions of production (Gallura's *caseus gallurensis*, or Arborea) or the cities (Cagliari's *caxeus salsus Calari*, Torres). This richness in specifications was very uncommon for this period.

Regarding specifically Sardinian cheeses, it is difficult to establish precisely the antecedents of *casu fràzigu* or *casu marzu* ("faded" or rotten cheese). It was a sheep's cheese or goat's cheese contaminated by the cheese fly (*Piophila casei*). Similar cheese varieties existed in Abruzzo, Apulia, and Calabria regions, but also in Northern Italy (Liguria, Emilia–Romagna, and Piedmont). Another original practice, also adopted in Sardinia, concerned the protection of the cheese with a clay coating during ripening. Contrary to the protocol for making *casu fràzigu* or *casu marzo*, this practice was useful to maintain the cheese uncontaminated by molds and at a controlled temperature.

As far as Sicily is concerned, the Messina's custom tariff of 1355 mentioned numerous varieties: fresh *caciocavallo*, the fresh cheese with or without salt, cooked cheese (*caseum scaldatum*), and another variety known as *tumazzu* from Alcamo (Giagnacovo 2007). One of the most important destinations for the export of these cheeses was certainly Venice. Here, the round pecorino and the cow's milk *caciocavallo* both produced in Sicily arrived by sea (Faugeron 2014). Also in Tuscany, there were reports on the *cicilianus* cheese and on a cheese from Messina (*caseus missanensis seu albus*) (Galoppini 2006; Lanconelli 2011). Over the recent years, there have been historically questionable efforts to link certain current varieties of Sicilian cheese with those of antiquity. The report of a minor tax exemption, enjoyed by the Viceroy Ugo Moncada in 1515 to export 200 *cantari* of cheese to Africa, has led to instances in current marketing campaigns of instantly making Moncada's cheese, the present Ragusano!

2.4   The Fifteenth Century and Pantaleone da Confienza's *Summa lacticiniorum*

After the demographic crisis of the fourteenth century, a marked resumption in Italian population growth followed in the second half of the fifteenth century. An upward trend began which, in spite of instances of plague (1575, 1630), lasted for centuries. As a percentage, the Italian population grew less than the French or English, yet still managing to increase from 10 to 18 million in three centuries (sixteenth to eighteenth).

In Central and Southern Italy, the effects of this phenomenon were not particularly striking. On the contrary, the growth of the population had important repercussions in the Po Valley. Based on infrastructures created in the previous centuries, this demographic increase favored a profound transformation in agrarian management. In particular, the wild forests of the lowland were gradually cultivated. In Lombardy, and in some areas of the Veneto region, advancements in irrigation technology allowed a threefold increase in hay yield per year. This had major and positive effects integrating farmed livestock with bovine transhumance towards the Alpine pastures of Lombardy (Brescia, Bergamasco, Valtellina, and Canton Ticino), Venice, and Trentino.

Under these conditions, the production of cow's milk cheeses in Northern Italy gradually overtook that of sheep's milk cheeses, where cow's milk cheeses had existed since the thirteenth century. Incontrovertible proof was the description of *vacherini*, consumed at the Savoy court and produced on both sides of the Western Alps (Airoldi 1996). Nevertheless, to state that such varieties were similar to Fontina cheese is perhaps too ambitious. The practice of mixing milk from different species (e.g., 2/3 of sheep's milk and 1/3 of cow's milk) spread at this time. Immediately, consumers of high social level appreciated the cheeses made with this mixture of milks. Thus wrote Doctor Savonarola, referencing the opinion of the *gulosi* from the court of Estensi, when he wrote the *Libreto de tutte le cosse che se magnano* (Camporesi 1990).

In particular, the know-how on cheesemaking was markedly diffused. Cheesemakers from the Po Valley (in particular, from Bergamo) were the most active, spreading their knowledge in Alpine pastures, Veneto, Southern Trentino, and Piedmont. Such cheesemakers worked mainly during the summer, because of the advantage of pastures having better quality (Varanini 1990; Nada Patrone 1996). In the sixteenth to seventeenth centuries, the Alps probably resembled an immense dairy farm (Camporesi 1990).

Over time, agriculture and cattle breeding integrated into a true system of production, especially in the Po Valley. In the fifteenth century, the Gonzaga farm holdings already hosted several hundred cows, with similar occurring in the Piedmont and Venetian areas. In the seventeenth to eighteenth centuries, the Po Valley was deindustrialized considerably, and in Northern Italy, the woolen industry (obviously associated with sheep rearing) declined completely. On the contrary, agriculture and livestock grew and trended towards innovation. The improvement in water management was one of the most significant factors. In particular, it concerned the construction of the *marcite*. These consisted on terrain permanently irrigated, allowing the protection of grasses during the low winter temperatures, thereby also intensive forage production. At the beginning of the nineteenth century, Foscolo wrote in the *Sepolcri*, exaggerating somewhat, that the only wealth of eighteenth-century Lombard nobility was farming: "*cui solo è dolce il muggito dei buoi/che dagli antri abdüani e dal Ticino/lo fan d'ozi beato e di vivande*" ("whose only joy is the bellow of the herds/in the stables near the Adda and Ticino rivers/that make him happily idle and well fed").

Although the documentary sources increased in quantity and precision (sometimes there were also quantitative reports that allow statistical assessment), it is still not easy even for the fifteenth century, to define precisely the specific characteristics of individual cheeses. Nevertheless, there were significant traces of a very profound and significant cultural transformation, as well as elements of continuity with the past. From the cultural point of view, the great development of cheese production and commerce from previous centuries did not limit the doubts of medieval physicians and dietitians with respect to cheese. Certainly, the poor chemical and biological knowledge determined this diffidence, since: "the mysterious mechanisms of coagulation and fermentation were seen with suspicion" (Montanari 2011). Medical knowledge had linked the cheese to the "humor theory." The advice for the population was for prudent consumption. A recommendation from the famous medical school of Salerno suggested for centuries to eat cheese in small doses (*caseus est sanus, quem dat avara manus*). Nevertheless, also within the same medical school there were recommendations more favorable to cheese consumption (the *Regimen sanitatis Salerni*). Applying the aforementioned "humor theory," some dietary treatises also suggested the combination and succession of cheese with other foods. The *Post carnes caseum duces* recommended that after eating fish people should follow with a dry food such as walnuts, but after eating meat, the consumption of cheese (presumably ripened) was suggested (Pucci Donati 2016).

Therefore, discussions on cheese were very frequent in the treatises of the fifteenth century. In general, there was agreement and appreciation for the nutritional characteristics of fresh cheeses. Conversely, ripened and salted cheese (the most important from the economic point of view) had an unpleasant aspect and odor, and were considered very difficult to digest and likely to cause flatulence and constipation. The general opinion considered ripened and salted cheese as a meal for people inhabiting villages instead of cities, and a food for hard workers, soldiers, friars, and monks. Certainly, ripened cheese was defined as nutritious and energizing. So much so that the early sixteenth-century author Ercole Bentivoglio wrote "the lover will rest little with his partner, if he has eaten a good piece of ripened cheese at dinner" (Camporesi 1990). People also appreciated the *caseus putrefactus*, which, already in the thirteenth century, had been a gift for Amadeus V of Savoy (Nada Patrone 1996).

The intersection of the developing agrarian system and the cultural debate made the well-known fifteenth-century treatise by Pantaleone da Confienza, the Piedmont physician and adviser to Ludovic, Duke of Savoy, is of immeasurable value. This treatise was the *Summa lacticiniorum sive tractatus varii de butyro, de caseorum variarum gentium differentia et facultate*. It almost corresponds to a cheese encyclopedia, a collection of protocols for making butter and cheese varieties, describing cheeses from different populations and territories. The book was printed in 1477, enjoying great success even into the following centuries. This volume, more than any other treatise, changed the European culture of cheese.

The *Summa* concretized the work done by the physician, who dealt extensively with dietary issues and habits of healthy and sick people. The third and final sections of this treatise illustrates the cheese varieties suitable or contraindicated for the various *complexiones* of humans (choleric, phlegmatic, or melancholic people); the most appropriate cheeses depending on age (children, and mature and elder people); and the most suitable cheeses for those people who were healthy. Compared to the medical issues mentioned elsewhere (e.g., those from the famous Salerno school of medicine), the *Summa* had an intermediate position. Preferably, it suggested eating cheeses neither too aged nor too fresh, and sometimes provided specific therapeutic recommendations. The *Summa* also addressed the relationships between cheese and society. Pantaleone da Confienza clearly notes that elites (kings, counts, marquises, nobles; Montanari 2011) gladly ate cheese as a complement to various diets. This was an important worldwide novelty. Nevertheless, the author continued to share the opinion that cheese had not to be eaten every day, only when needed (*ad quottidianam casei commestionem impellit necessitas*).

For the society of the day, the *Summa* represented a cheese atlas. Pantaleone da Confienza had gained broad and practical experience, paying attention not only to the geographical distribution of cheesemaking but also to the overall attributes of the quality of cheese. He neglected several Mediterranean productions and excluded buffalo cheeses, *pasta filata* cheeses, blue cheeses, and putrefied cheeses (the latter considered to be corrupt and unhealthy). Having travelled long in the Alps, he had clear concepts borrowed from a European document named the "report on the fat compounds," and the related links with the nutritional aspects of cheese production. Apart from England (he appreciated the quality of the British cheeses exported to Anversa, Naso 2016), he considered of scarce quality the cheeses manufactured in several countries of the north of Europe such as Brittany, Flanders, Brabant, Hainaut, Artois, and Holland. This judgment relied on the widespread use of skimmed milk. The assessment for Italian and French cheeses was very different because of the widespread use of olive oil and lard as cooking ingredients, and because of the use of whole milk for cheesemaking (Naso 1999). Pantaleone da Confienza considered cow's cheese of supreme importance, virtually consecrating the prestige of *parmigiano* or *piacentino* cheese, together with recognizing the excellence of *marzolino* cheese (also exported to France and consumed by him). Of course, he had a special regard for cheeses from his own region, Piedmont. He knew the *robiole* from the southern of Piedmont well, as well as the cheeses from the Aosta, Susa, and Lanzo Valleys, and the *savoiardi* (Maurienne, Tarantasia, and Bresse).

Although not extensive, the *Summa* also provides technical information on cheesemaking. In contrast to what Pier de' Crescenzi had done in the early fourteenth century, the *Summa* first described cheese technology without repeating the information derived from antiquity. Pantaleone da Confienza was aware of the characteristics of rennet (origin, freshness, and quantity) and its effect on cheese quality. He insisted on curd pressing technology to ensure a compact shape, without sponginess or holes, he also cited tools for making Ricotta (Fig. 2.6). Further, he noted the importance of the seasons for cheesemaking, aware of the varying quality of milk throughout the year. Also considered was the importance of the relationship between crust dryness and cheese spoilage. Additionally, he emphasized the importance of adding salt, thus linking the presence of salt to the cheese shape and quality. Pantaleone da Confienza suggested 3–4 years of salting for *parmigiano/piacentino*, or 6–8 months for *robiole*, depending on the characteristics of the ripening room (dry, heated, or aerated). He also observed the link between the characteristics of a cheese and its *accidentalis form*, referring to its "geometrical" classifications. There were cylindrical (*rotundi et oblongi ad quantitatem semicubiti,* circular and oblong, like a forearm) such as the *marzolini*, spherical (*quasi ex toto rotundi*); parallelepiped; truncated cone; triangular prism; or unshaped. The treatise further notes how the dimensions of these cheeses varied over time. During the fifteenth century, the average weight of a *parmigiano* round increased to 18–20 kg and above, to even over 30 kg (Giagnacovo 2007).

Concluding, Pantaleone da Confienza did not neglect the importance of the human aspect in this process, which in his view was always decisive. He even dictated the ideal physical attributes of the cheesemaker. Their hands had to be soft and delicate, for optimally shaping the cheese. Hands too warm caused *cavernositas* (presence of holes) during cheese manipulation. This was one of the most feared defects: *nam videmos caseos oculatos sive spongiosos vel cavernosos comuniter ab omnibus detestari* (indeed, the cheeses with eyes, spongy and cavernous are detested by everybody) (Naso 1990, 1996) (Fig. 2.7).

Fig. 2.7 The toponyms mentioned in the text. Cities: indicated with a circle. Districts and valleys: in a box. Single production locations: indicated with a black dot. The borders of the regions are the modern-day ones

References

Airoldi P (1996) La tavola del conte. Spese per il cibo alla corte di Filippo I di Savoia. In: Cumba R, Nada Patrone AM, Naso I (eds) La mensa del Principe. Cucina e regimi alimentari nelle corti sabaude (XIII-XV sec.). Società studi storici, Alba-Cuneo, pp 9–42

Amiotti G (2011) Produzione, commercio e uso del formaggio nell'antica Roma. In: Archetti G, Baronio A (eds) La civiltà del latte. Fonti, simboli e prodotti dal tardo antico al Novecento, Atti dell'incontro nazionale di studio. Fondazione Civiltà Bresciana, Brescia, pp 15–21

Archetti G (2011) Vas optimo lacte plenum. Latte e formaggio nel mondo monastico. In: Archetti G, Baronio A (eds) La civiltà del latte. Fonti, simboli e prodotti dal tardo antico al Novecento, Atti dell'incontro nazionale di studio. Fondazione Civiltà Bresciana, Brescia, pp 249–278

Aymard M (1983) Conoscenza ed incertezza sui consumi di formaggio in Francia e in Italia tra il XIV e il XVII secolo. In: Il caseario. Un archetipo alimentare: il latte e le sue metamorfosi, Atti del convegno internazionale, Bologna 14–15 aprile 1983. CLUEB, Bologna, pp 27–30

Basso E (2011) Circolazione e commercio Dei prodotti caseari nel Mediterraneo (secoli XIII–XV). In: Archetti G, Baronio A (eds) La civiltà del latte. Fonti, simboli e prodotti dal tardo antico al Novecento, Atti dell'incontro nazionale di studio. Fondazione Civiltà Bresciana, Brescia, pp 79–101

Beccaria GL (2007) Tra le pieghe delle parole. Lingua storia cultura, Einaudi, Torino

Bellini R (2011) Il latte e il formaggio nei testi penitenziali e nelle fonti canonistiche. In: Archetti G, Baronio A (eds) La civiltà del latte. Fonti, simboli e prodotti dal tardo antico al Novecento, Atti dell'incontro nazionale di studio. Fondazione Civiltà Bresciana, Brescia, pp 309–366

Bonazza C (2011) Economia e lavorazione Dei prodotti lattiero-caseari negli statuti e carte di regola tardo-medievali. In: Archetti G, Baronio A (eds) La civiltà del latte. Fonti, simboli e prodotti dal tardo antico al Novecento, Atti dell'incontro nazionale di studio. Fondazione Civiltà Bresciana, Brescia, pp 499–540

Camporesi P (1990) Certosini e marzolini: l'iter casearium di Pantaleone da Confienza nell'Europa dei latticini. In: La miniera del mondo. Artieri, inventori, impostori, Garzanti, Milano, pp 89–117

Carocci S, Vendittelli M (2004) L'origine della Campagna Romana. Casali, Castelli e villaggi nel XII e XIII secolo. Società romana di Storia patria, Roma

Comba R (2011) Alpeggi, saperi naturalistici e caseari, "natura" Dei formaggi. Qualche riflessione storiografica su un secolo di ricerche. In: Archetti G, Baronio A (eds) La civiltà del latte. Fonti, simboli e prodotti dal tardo antico al Novecento, Atti dell'incontro nazionale di studio. Fondazione Civiltà Bresciana, Brescia, pp 23–40

Di Muro A (2011) La terra dove scorre latte e miele. Per una storia della produzione di latte e formaggio nel Mezzogiorno medievale. In: Archetti G, Baronio A (eds) La civiltà del latte. Fonti, simboli e prodotti dal tardo antico al Novecento, Atti dell'incontro nazionale di studio. Fondazione Civiltà Bresciana, Brescia, pp 55–78

Faugeron F (2014) Nourrir la ville. Ravitaillement, marchés et métiers de l'alimentation à Venise dans les derniers siècles du Moyen Âge. Ècole Francaise de Rome, Roma

Galoppini L (2006) Produzione e commercio dei formaggi nella Toscana del medioevo. In: Bollettino della Accademia degli Euteleti della città di San Miniato, pp 407–435

Giagnacovo M (2007) Formaggi in tavola. Commercio e consumo del formaggio nel basso Medioevo. Un contributo dell'Archivio Datini di Prato. Aracne, Roma

Greci R (2011) Il commercio di generi alimentari. Norme corporative e potere pubblico. In: Archetti G, Baronio A (eds) La civiltà del latte. Fonti, simboli e prodotti dal tardo antico al Novecento, Atti dell'incontro nazionale di studio. Fondazione Civiltà Bresciana, Brescia, pp 541–653

Lanconelli A (2011) Il formaggio nel medioevo. Storiografia sull'Italia centrale. In: Archetti G, Baronio A (eds) La civiltà del latte. Fonti, simboli e prodotti dal tardo antico al Novecento, Atti dell'incontro nazionale di studio. Fondazione Civiltà Bresciana, Brescia, pp 41–53

Lombardo ML (1983) La dogana minuta a Roa nel primo Quattrocento. Aspetti istituzionali, sociali, economici, Il Centro di Ricerca—Fonti e studi del Corpus membranarum italicarum, Roma

Maire Vigueur JC (1974) Les "casali" des églises romaines à la fin du Moyen Âge. Mélanges de l'École française de Rome (1348–1428) 86:63–136

Maire Vigueur JC (1981) Les pâturages de l'Église et la Douane du bétail dans la Province du Patrimonio (XIVe-XVesiècles) Roma

Martin JM (2007) Les débuts de la transhumance. Économie et habitat en Capitanata. Bollettino dell'Istituto storico italiano per il medioevo, pp 117–137

Montanari M (2005) Ruolo del latte e dei formaggi nel medioevo. www.mondimedievali.net/pre-testi/montanari.htm

Montanari M (2008) Il formaggio con le pere. La storia in un proverbio. Laterza, Roma-Bari

Montanari M (2011) Prodotti e simboli alimentari. Latte e formaggio tra economia e cultura. In: Archetti G, Baronio A (eds) La civiltà del latte. Fonti, simboli e prodotti dal tardo antico al Novecento, Atti dell'incontro nazionale di studio. Fondazione Civiltà Bresciana, Brescia, pp 3–15

Muzzarelli MG (2015) Margherita Datini e Alessandra Macinghi Strozzi spediscono, ricevono e smistano cibi. Progressus. Rivista di storia, scrittura e società 2:33–53

Nada Patrone AM (1996) Caseus est sanus quod dat avara manus. In: Comba R, Dal Verme A, Naso I (eds) Il consumo del formaggio dal XII al XVII secolo, in Greggi, mandrie e pastori nelle Alpi occidentali (secoli XII–XX). Società per gli studi storici, archeologici e artistici della Provincia di Cuneo, Cuneo-Rocca de' Baldi, pp 97–122

Naso I (1990) Formaggi del medioevo. La summa lacticiniorum di Pantaleone da Confienza. Il Segnalibro, Torino

Naso I (1996) Una risorsa dell'allevamento. Aspetti tecnici e culturali della lavorazione del latte nel Quattrocento. In: Comba R, Dal Verme A, Naso I (eds) Greggi, mandrie e pastori nelle Alpi occidentali (secoli XII–XX). Società per gli studi storici, archeologici e artistici della Provincia di Cuneo, Cuneo-Rocca de' Baldi, pp 125–148

Naso I (1999) La cultura del cibo. Alimentazione, dietetica, cucina nel basso medioevo. Paravia-Scriptorium, Torino

Naso I (2000) Università e sapere medico nel Quattrocento. Pantaleone da Confienza e le sue opere. Società per gli studi storici, archeologici e artistici della Provincia di Cuneo, Cuneo

Naso I (2016) Produzione e consumo dei latticini in età premoderna, fra teorie mediche e pratiche quotidiane. In: Agricoltura, alimentazione e sostenibilità. Scienza attiva, edn 2015/2016, Torino pp 1–5

Parenti S (2011) Il formaggio nella liturgia e nelle consuetudini monastiche bizantine. In: Archetti G, Baronio A (eds) La civiltà del latte. Fonti, simboli e prodotti dal tardo antico al Novecento, Atti dell'incontro nazionale di studio. Fondazione Civiltà Bresciana, Brescia, pp 279–308

Pession A (ed) (2015) Un esempio dell'amministrazione medievale sabauda: il primo conto della castellania di quart e Oyace (1377–1378). Regione Autonoma Valle d'Aosta-Assessorato Istruzione e Cultura-Archivio storico regionale, Aosta

Pucci Donati F (2016) Approvvigionamento distribuzione e consumo in una città medievale. Il mercato del pesce a Bologna (secoli XIII–XV). Centro italiano di studi sull'alto medioevo, Spoleto

Varanini GM (1990) Una montagna per la città. Alpeggio e allevamento nei Lessini veronesi nel Medioevo (secoli IX–XV). In: Berni P, Sauro U, Varanini GM (eds) Gli alti pascoli dei Lessini. Natura storia cultura. La Grafica, Vago di Lavagno (Verona), pp 13–106

Vigolo MT (1990) Termini di interesse storico relative ad alcuni tipi di formaggio in uso nella Lessinia veronese, attestati in documenti medievali. In: Berni P, Sauro U, Varanini GM (eds) Gli alti pascoli dei Lessini. Natura storia cultura. La Grafica, Vago di Lavagno (Verona), pp 30–31

Violante F (2009) Il re, il contadino, il pastore. La grande masseria di Lucera e la dogana delle pecore di Foggia tra XV e XVI secolo. Edipuglia, Bari

Chapter 3
Cheese: An Overview

3.1 Introduction

Cheese is the most diverse group of dairy products and is, arguably, the most academically interesting and challenging. While most dairy products, if properly manufactured and stored, are biologically, biochemically, chemically, and physically very stable, cheeses are biologically and biochemically dynamic, and are inherently unstable. Throughout manufacture and ripening, cheese production represents a finely orchestrated series of consecutive and concomitant biochemical events, which, if synchronized and balanced, lead to products with highly desirable aromas and flavors but when unbalanced, result in off-flavors and odors.

A further important aspect of cheese is the range of scientific disciplines involved: study of cheese manufacture and ripening involves the chemistry and biochemistry of milk constituents, fractionation and chemical characterization of cheese constituents, microbiology, enzymology, molecular genetics, flavor chemistry, nutrition, toxicology, rheology, and chemical engineering. A voluminous scientific and technological literature has accumulated, including several books (see Fox et al. 2017; Tunick 2014; McSweeney et al. 2017). In addition, there are numerous encyclopedias or pictorial books, e.g., Mair-Waldburg (1974), Harbutt (1999, 2002), and Donnelly (2016). There are also a number of country-specific or variety-specific books (see McSweeney et al. 2017; Mucchetti and Neviani 2006).

3.2 Outline of Cheese Manufacture

This chapter provides an overview of cheese manufacture to familiarize the reader with the terminology used in later chapters. The principle of cheese manufacture involves coagulation of the caseins by acidification to pH ca. 4.6, for ca. 20% of all cheeses, or by specific, limited, proteolysis, for ca. 75% of all cheeses, or by

© Springer International Publishing AG, part of Springer Nature 2018
M. Gobbetti et al., *The cheeses of Italy: Science and Technology*, https://doi.org/10.1007/978-3-319-89854-4_3

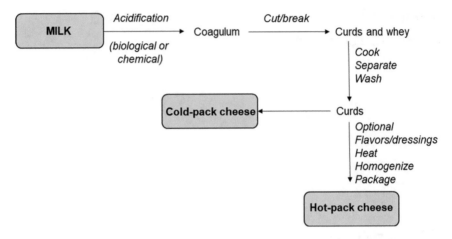

Fig. 3.1 Outline flowsheet for the manufacturer of acid-curd cheese

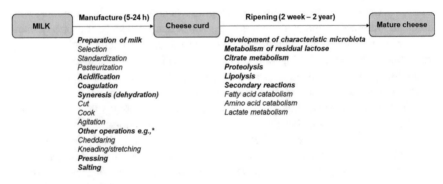

*e.g., bactofugation, microfiltration, addition of colour (annato)

Fig. 3.2 Outline flowsheet for the manufacture of rennet-coagulated cheese

acidification of a mixture of rennet whey and skimmed or whole milk to ca. pH 5.2 at ca. 90 °C. If fat is present, it is occluded in the coagulated protein.

Almost all acid-coagulated, acid–heat-coagulated, and a little rennet-coagulated cheeses are consumed fresh. The flavor, texture, and appearance of the cheese are in their final form at the end of curd production and the curds do not undergo to a period of maturation/ripening. Figure 3.1 summarizes the production of acid-coagulated cheeses.

Ripening is common for rennet-coagulated cheeses and their production may be subdivided into two well-defined phases, manufacture and ripening, both of which involve a number of processes (Fig. 3.2). The manufacturing phase concerns those operations performed during the first 24 h, although some of these operations, e.g., salting and dehydration, may continue over a longer period. Although the manufacturing protocol for individual varieties differs in detail, the basic steps are common to most varieties; these are: preparation of milk, acidification, coagulation, dehydration (cutting the coagulum, cooking, stirring, pressing, salting, and other operations that promote gel syneresis), separation of curds and whey, shaping (molding and pressing), and salting.

Cheese manufacture is essentially a dehydration process in which the fat and casein of milk are concentrated 6- to 12-fold, depending on the variety. The degree of dehydration is regulated by the extent and combination of the above five operations, in addition to the chemical composition of the milk. The levels of moisture and salt, the pH, and the cheese microbiota regulate and control the biochemical changes that occur during ripening and hence determine the flavor, aroma, texture, and functionality of the finished cheese. The manufacturing steps determine the nature and quality of the finished cheese, but it is during the ripening phase that the characteristic flavor and texture of the individual cheese varieties develop.

3.3   Selection and Pre-treatment of Cheese Milk

Cheese manufacture commences with the selection of milk of high microbiological and chemical quality. All aspects of cheese curd production (rennet coagulation, gel firmness, syneresis) are affected by the chemical composition of the milk, especially the concentrations of casein, calcium (Ca^{2+}), and pH. The chemistry of milk and milk constituents is described in Walstra et al. (2005) and Fox et al. (2017).

The milk for many Italian cheese varieties is not cooled before cheesemaking (see Chaps. 5 and 6), but in northern Europe, North America, Australia, and New Zealand the milk is cooled to ca. 4 °C immediately after milking, being held at about this temperature for several days on the farm and at the factory. Apart from being selective for the development of an undesirable psychrotroph microbiota, cold storage causes physicochemical changes (e.g., a shift in calcium phosphate equilibrium and dissociation of some micellar caseins), which have undesirable effects on the cheesemaking properties of milk. These changes are reversed on heating, e.g., at 50 °C for 10–20 min or on HTST (High Temperature Short Time) pasteurization and hence are of no practical significance. However, cold storage after heat treatment aggravates the adverse effects of heating on the rennet coagulation of milk, an effect known as rennet hysteresis.

The composition of most cheeses falls within certain ranges, sometimes with legal status. The most important compositional factors are fat-in-dry matter (FDM), moisture in non-fat substances (MNFS; which is, in effect, the ratio of moisture to protein), moisture, salt (best expressed as salt-in-moisture, S/M), and pH. The composition of the cheese milk and extent of syneresis, respectively, determine the values of FDM and MNFS. The composition of milk is adjusted to give the desired values of fat and protein. Previously, the ratio between fat and proteins was altered by natural creaming (which is still used for Parmigiano Reggiano, Grana Padano, and some other Italian cheeses) or by the addition of cream or skim milk. It is now possible, and commercially practiced, to adjust the concentrations, as well as the ratio, of fat and casein in the milk by adjusting the casein content by low-concentration factor ultrafiltration (UF). The use of UF also increases the capacity of the cheese plant.

Owing to the importance of Ca^{2+} in various aspects of cheese manufacture and quality, it is common practice to supplement cheese milk with $CaCl_2$. Nevertheless, this practice is rather unusual in Italian cheesemaking, and it is prohibited for all the PDO (Protected Denomination of Origin) cheese production. The pH is also a critical factor in cheesemaking, and since the pH of milk varies (e.g., increases with advancing lactation and during mastitic infection), it is recommended that the pH should be standardized, preferably using the acidogen, gluconic acid-delta-lactone.

Milk from a healthy mammary gland is sterile but becomes contaminated with bacteria during milking and afterwards. With a good machine milking operation and healthy animals, it is possible to get a total bacterial count (TBC) of less than 10^4–10^5 cfu/mL. Contamination is normally much higher with hand milking. If the milk is not cooled quickly, the indigenous microorganisms will grow rapidly and in a warm climate milk will sour quickly, perhaps within 2 h.

The adventitious microbiota of milk is normally heterogeneous; some of these microorganisms, especially the lactic acid bacteria, may be beneficial. Previously, and still for some minor artisanal cheeses, the autochthonous lactic acid bacteria are responsible for acid production but selected cultures (starters) are used for acidification in most cases. Non-starter lactic acid bacteria (NSLAB) probably contribute positively to the ripening of raw milk cheese but they are variable and uncontrolled and may be responsible for some of the variability in raw milk cheese (see Chap. 5). The milk for most Italian cheese varieties is not pasteurized (see Chap. 6) and the NSLAB are critically important. For many cheesemaking operations, other than in Italy, it is normal practice to kill the NSLAB by pasteurization (although this is not the primary objective of pasteurization). There is increasing interest in inoculating pasteurized milk with selected lactobacilli as an adjunct culture.

Some members of the adventitious microbiota are undesirable. The most important are a number of pathogens, the killing of which is the primary objective of pasteurization. Raw milk may also contain spoilage microorganisms, e.g., coliforms, psychrotrophs, and *Clostridium tyrobutyricum*. Growth of this spore-forming organism during the ripening of most cheese varieties results in a defect known as late gas blowing, caused by the anaerobic metabolism of lactate to butyrate and H_2. Contamination with *Cl. tyrobutyricum* is minimized by good on-farm hygiene, contaminants may be removed by bactofugation or microfiltration, or the use of $NaNO_3$ or lysozyme inhibits their growth.

Cheese milk must be free from antibiotics, which totally, or partially, inhibit the starter bacteria; delayed acidification results in an abnormal composition of the microbiota and, consequently, in flavor and textural defects and perhaps very significantly in the growth of harmful, pathogenic or food poisoning microorganisms. Raw milk is still used in both industrial and farmhouse cheesemaking, especially in Italy and other southern European countries, but elsewhere, most cheese milk is pasteurized, usually immediately before use. Pasteurization alters the indigenous microbiota and facilitates the manufacture of cheese of more uniform quality, but unless due care is exercised; it may damage the rennet coagulability and curd-forming properties of the milk. Even when milk is properly pasteurized, the resulting cheese develops a less intense flavor and ripens more slowly than raw milk cheese. Several heat-induced changes, e.g., killing of autochthonous microorganisms, inactivation of indigenous milk enzymes, and limited denaturation of whey proteins and their interaction with micellar κ-casein, could be responsible for these changes. The relative contribution of these factors to the differences between cheeses made from raw or pasteurized milk has been an active area of research over the recent years (Fox et al. 2017).

A number of approaches may render cheese milk free from its autochthonous microbiota or inhibit the growth of NSLAB in order to study their contribution to ripening (see Chap. 5). NSLAB have been removed physically from raw skim milk by microfiltration, contamination from the environment has been minimized by manufacturing cheeses under strictly controlled microbiological conditions, ripening cheese at a low (ca. 1 °C) temperature to reduce/prevent the growth of NSLAB and the use of antibiotics to inhibit the growth of NSLAB. Attempts have been made to mimic the NSLAB microbiota of raw milk cheese by adding selected strains of NSLAB to pasteurized cheese milk or by inoculating pasteurized cheese milk with raw milk (by blending a low level, e.g., 1%, of raw milk with pasteurized milk). The results of these studies suggest that heat-induced changes to the microbiota of raw milk are the principal cause of the differences between raw and pasteurized milk cheeses. However, denaturation of certain indigenous enzymes, particularly lipoprotein lipase, may contribute to the observed differences.

Pasteurization of cheese milk minimizes the risk of cheese serving as a vector for food poisoning and pathogenic microorganisms, so that even high-quality raw milk may be unacceptable for cheese manufacture. In addition to rendering milk safe from a public health viewpoint, pasteurization renders good quality raw milk almost free of bacteria and improves the consistency of cheese. Pasteurization of milk is essential for the production of cheese of consistent quality in the large, highly mechanized factories that are common today. Although more consistent than cheese made from raw milk, it is less highly flavored. To increase the intensity of the flavor of cheese made from pasteurized milk, it is becoming increasingly common to inoculate pasteurized milk with selected organisms, usually lactobacilli, isolated from good quality raw milk cheese. Thermization (65 °C for 15 s) of cheese milk on arrival at the factory is common or normal practice in some countries, including Italy. The objective is to control psychrotrophs and the milk is normally pasteurized before cheesemaking. Microfiltration and bactofugation may be used to remove spores from milk to avoid the defect known as late gas blowing. Well-made cheese is a rather hostile environment and pathogenic bacteria die off in cheese during ripening; it is assumed that pathogens die off within 60 days but this is not certain in several cases. In this technological context, Italy may be a representative exception. As shown by the very long tradition of many long-ripened Italian cheeses (e.g., Parmigiano Reggiano, Grana Padano, and several other varieties made from both cow's milk and ewe's milk), the use of the raw milk cheese is safe like the pasteurized milk because of the capacity of the prolonged maturation to safely prevent microbiological risks.

Milk contains somatic cells (SC; leucocytes; lymphocytes, phagocytes, mammary gland epithelial cells) which contain many hydrolytic enzymes. Mastitic milk may contain more than 10^6 SC/mL and even good quality milk probably contains 100,000 SC/mL. A high somatic cell count will reduce the yield of cheese and affect its quality. The effects of high SC content on cheese yield and quality have been studied extensively but the effects of low/normal SC content have received little attention. Somatic cells may be removed by ultrafiltration but this is not commonly practiced and the benefits have not been quantified.

Not more than 75% of the total protein in milk is recovered in rennet-coagulated cheeses. Obviously, a considerable economic advantage would accrue if some or all of the whey proteins could be incorporated into the cheese. Ultrafiltration offers a means for accomplishing this, with considerable success in the case of soft or semi-soft cheeses, especially Crescenza, Feta, and Quarg, but not for hard and semi-hard varieties. An alternative approach is to heat-denature the whey proteins (e.g., 90 °C for 1 min) to induce their interaction with the casein micelles. Normally, such severe heat treatments are detrimental to the renneting properties of milk but the effects may be offset by acidification or supplementation with calcium. In the authors' experience, yield increases of up to 8% may be achieved by this approach, while retaining acceptable quality. However, to the authors' knowledge, the technique is not used commercially except for Quarg, e.g., the thermoQuarg process (McSweeney et al. 2017).

Recovery of whey proteins by microfiltration and commercialization as whey protein concentrate or whey protein isolate may be more economical than their incorporation into cheese.

3.4 Acidification

One of the basic operations in the manufacture of all cheese varieties is a progressive acidification throughout the manufacturing stage, e.g., up to 24 h, and, for some varieties, during the early stages of ripening also, e.g., acidification commences before and transcends the other manufacturing operations. Acidification is normally achieved by in situ production of lactic acid, although acid (citric acid) or acidogen (usually gluconic acid-delta-lactone) is now added to milk to directly acidify curd for some varieties (e.g., Mozzarella/Pizza, UF Feta-type, and Cottage cheese). Until relatively recently, and still in some cases, especially for artisanal varieties, the autochthonous microbiota of milk was relied upon for acid production. Since this was probably a mixed biota, the rate of acid production was unpredictable and the growth of undesirable bacteria led to the production of gas and off-flavors. It is now almost universal practice in industrial cheesemaking to add a culture (starter) of selected lactic acid bacteria to raw or pasteurized cheese milk to achieve a uniform and predictable rate of acid production. Natural milk or whey cultures are widely used for Italian cheeses (see Chaps. 5 and 6). Usually, the microbial composition of natural milk or whey cultures depends on the temperature used for their manufacture, which selects thermophilic or mesophilic lactic acid bacteria. A culture containing *Streptococcus thermophilus* and a *Lactobacillus* spp. (*Lactobacillus delbrueckii* subsp. *bulgaricus, Lactobacillus delbrueckii* subsp. *lactis,* or *Lactobacillus helveticus*) or a *Lactobacillus* culture alone is used for many Italian varieties, including those that are cooked to a temperature higher than 40 °C. Unlike the common practice of cheesemaking in other countries, the use of thermophilic starter cultures is widespread also for Italian cheese varieties that are not cooked or cooked to less than 40 °C.

Originally, and until relatively recently, acidification of cheese depended on the autochthonous microbiota of the milk, especially lactic acid bacteria. This biota is variable and unpredictable. The acidification of many artisanal cheeses still depends on the indigenous microbiota. However, for consistency, the milk for all factory-produced cheese is inoculated with a culture (starter) of lactic acid bacteria.

The earliest form of starter was a "back slopping" culture: a sample of whey from one day's cheesemaking was incubated overnight and used as a starter culture on the following day. Such starters are largely used for some high- and low-cooked Italian cheese varieties (e.g., Parmigiano Reggiano, Grana Padano, and other varieties). Incubation of hot whey is selective for thermophilic microorganisms, and although back slopping cultures are heterogeneous, they work very well if managed carefully.

Originally, and still for many varieties, mixed-strain mesophilic starters were used for low-cooked cheese. Because the bacterial strains in these starters may be phage-related (e.g., subject to infection by a single strain of bacteriophage) and also because the strains in the mixture may be incompatible, thereby leading to the dominance of one or a few strains, the rate of acid production by mixed-strain starters is variable and unpredictable, even when the utmost care in their selection and handling is exercised. To overcome these problems, single-strain mesophilic starters were introduced in New Zealand around 1935. Unfortunately, many of the fast acid-producing, single-strain starters produced bitter cheese. This problem was resolved by using selected pairs of fast and slow acid producers. Defined-strain mesophilic starters are widely used in many countries, frequently consisting of a combination of 2–6 selected, phage-unrelated strains, which give a very reproducible rate of acid production, if properly selected and maintained. The use of defined-strain thermophilic starters is fairly common.

In-house propagation of starters is laborious and requires skilled technicians. An alternative is to use "direct-to-vat" concentrated starter (DVS) cultures. These are prepared and marketed by specialized culture producers, concentrated and frozen. The cheesemaker simply adds the frozen concentrated culture to the milk in the cheese vat, at a level of perhaps 1 L per 10,000 L of milk.

Acid production at the appropriate rate and time is the key step in the manufacture of good quality cheese (excluding the enzymatic coagulation of the milk, which is a *sine qua non* for rennet-coagulated cheese varieties). Acid production affects several aspects of cheese manufacture:

– Coagulant activity during coagulation.

- Denaturation and retention of the coagulant in the curd during manufacture and hence the level of residual coagulant in the curd; this influences the rate of proteolysis during ripening, and may affect cheese quality.
- Strength of the coagulum, which affects the loss of fines in the whey and influences cheese yield.
- Gel syneresis, which affects cheese moisture and hence regulates the growth of bacteria and the activity of enzymes in the cheese; consequently, it strongly influences the rate and pattern of ripening and the quality of the mature cheese.
- The rate of acidification determines the extent of dissolution of colloidal calcium phosphate which modifies the susceptibility of the caseins to proteolysis during manufacture, influences the rheological properties of the cheese, e.g., compare the texture of Emmental, Gouda, Grana Padano, Cheddar, and Cheshire cheese, and determines the meltability and stretchability of cheese curd (e.g., Mozzarella and Pizza cheese).
- Acidification controls the growth of many species of bacteria in cheese, especially pathogenic, food poisoning, and gas-producing microorganisms. Properly made cheese is a very safe product from the public health viewpoint. In addition to producing acid, many starter bacteria produce bacteriocins that restrict or inhibit the growth of non-starter microorganisms.

Mesophilic *Lactococcus* spp. are capable of reducing the pH of cheese to ca. 4.6 and *Lactobacillus* spp. to a somewhat lower value, perhaps 3.8. The natural ultimate pH of cheese curd falls within the range 4.6–5.1. However, the period required to attain the ultimate pH varies from 5 h for Cheddar to 6–12 h for blue, Dutch, Swiss, and Italian varieties. The differences arise from the amount of starter added to the milk (0.2–5%), the cooking temperature, and the rate of subsequent cooling of the curd.

The level and method of salting have a major influence on pH changes in cheese. The concentration of NaCl in cheese (commonly 0.7–4%, e.g., 2–10% salt in the moisture phase) is sufficient to halt the growth of starter bacteria. Some varieties, mostly of British origin, are salted by mixing dry salt with the curd towards the end of manufacture and hence the pH of curd for these varieties must be close to (ca. pH 5.4) the ultimate value (pH 5.1) at salting. However, most varieties are salted after molding by immersion in brine or by application of dry salt on the surface; salt diffusion in cheese moisture is a slow process and there is ample time for the pH to decrease to 5.0 before the salt concentration becomes inhibitory. The pH of the curd for most cheese varieties, e.g., most Italian varieties, Emmental, Gouda, and blue, is 6.2–6.5 at molding and pressing but decreases to 5 during or shortly after pressing and before salting. The significance of various aspects of the concentration and distribution of NaCl in cheese were discussed by Guinee and Fox (2017).

In a few special cases, e.g., Domiati, a high level of NaCl (10–12%) is added to the cheese milk, traditionally to control the growth of the indigenous microbiota. This concentration of NaCl has a major influence, not only on acid development but also on rennet coagulation, gel strength, and syneresis (McSweeney et al. 2017).

3.5 Coagulation

The essential characteristic step in the manufacture of all cheese varieties is coagulation of the caseins to form a gel, which entraps the fat, if present. Coagulation may be achieved by (1) limited proteolysis by selected proteinases (rennets); (2) acidification to pH 4.6, and (3) acidification to pH about 5.2 in combination with heating to 90 °C.

Most varieties of cheese (ca. 75% of total production) are produced by enzymatic (rennet)-induced coagulation. With a few exceptions (e.g., Serra da Estrêla—Portugal—in which acid proteinases from the flowers of the cardoon thistle, *Cynara cardunculus*, are used), acid (aspartyl) proteinases of animal or fungal origin are used. Rennet from the stomachs of young animals (calves, kids, lambs, buffalo) was used traditionally. The principal enzyme in rennet prepared from young animal stomachs is chymosin (95% of total milk-clotting activity), with a little pepsin; as the animal ages, the secretion of chymosin decreases and that of pepsin increases. However, a limited supply of such rennets, due to the birth of fewer calves and an increasing trend in many countries to slaughter calves at an older age than previously, concomitant with a worldwide increase in cheese production, has occurred. This has led to a shortage of calf rennet and consequently rennet substitutes (e.g., bovine or porcine pepsins and less frequently chicken pepsin, and the acid proteinases from *Rhizomucor miehei* and less frequently from *Rhizomucor pusillus* or *Cryphonectria parasitica*) are used widely for cheese manufacture with more or less satisfactory results. The calf chymosin gene has been cloned in *Kluyveromyces lactis, Escherichia coli*, and *Aspergillus niger*, and chymosin from these organisms is now widely used. The gene for camel chymosin also has been cloned. Camel chymosin has very good milk-clotting activity but lower general proteolytic activity than calf chymosin and is being used commercially in certain situations. For reviews on rennet substitutes, see Jaros and Rohm (2017). The molecular and enzymatic properties of calf chymosin and other acid proteinases used as rennets were reviewed by Uniacke-Lowe and Fox (2017).

In 1942, N.J. Berridge demonstrated that the rennet-catalyzed coagulation of milk occurs in two phases: a primary enzymatic phase and a secondary non-enzymatic phase. The primary phase has a temperature coefficient (Q_{10}) of ca. 2 and occurs in the range 0–50 °C, while the secondary phase has a Q_{10} of ca. 16 and occurs very slowly at a temperature lower than 18 °C; aggregation occurs slowly at a low temperature but gelation does not occur (Bansal et al. 2007). The two phases can thus be separated readily by performing the primary phase at a low temperature, e.g., 10 °C; when cold-renneted milk is warmed, coagulation occurs very quickly. Cold renneting, followed by rapid warming, is the principle of attempts to develop methods for the continuous coagulation of milk but such approaches have not been successful commercially. Normally, the two phases of rennet coagulation overlap, the magnitude of overlap being quite extensive at low pH or high temperature and in milk concentrated by ultrafiltration.

The primary phase of rennet-induced coagulation was recognized, in general terms during the 1880s, by Olav Hammersten, who reported the formation of small peptides during renneting. The coagulation of milk attracted much interest during the first half of the twentieth century but an explanation of the process had to await the isolation of κ-casein, by Waugh and von Hippel (1956). κ-Casein is the only milk protein hydrolyzed during the primary phase of rennet action. Only one peptide bond, Phe_{105}-Met_{106}, is hydrolyzed, resulting in the release of the hydrophilic C-terminal segment of κ-casein [the (caseino)macropeptides, some of which are glycosylated]. The unique sensitivity of the Phe-Met bond of κ-casein has been the subject of considerable research (Fox et al. 2017).

The visual coagulation of milk is only the start of the gelation process, which continues for a considerable period thereafter. Although these post-coagulation changes determine many of the critical cheesemaking properties of the gel, e.g., curd tension and syneretic properties, it is the least well understood phase of the cheesemaking process. The recent literature on aspects of the post-visual coagulation phase was reviewed by Fagan et al. (2017).

3.6 Post-coagulation Operations

A rennet-coagulated milk gel is quite stable under quiescent conditions but if it is cut or broken, it syneresis rapidly, expelling whey. The rate and extent of syneresis are influenced, *inter alia*, by how finely the coagulum is cut (cutting into small pieces promotes syneresis; the coagulum for high-moisture cheeses is not cut but is ladled into molds), milk composition, especially the concentrations of Ca^{2+} and casein (up to a point; at a very high concentration of casein, the gel is very stiff and does not synerese well), pH, cooking temperature, rate of stirring of the curd–whey mixture, and time (Fagan et al. 2017). The composition of the finished cheese is strongly determined by the extent of syneresis, and since this is under the control of the cheesemaker, the differentiation of cheese varieties really begins at this stage, although the composition of cheese milk, the amount and type of starter, and the amount and type of rennet are also significant in this regard.

The temperature to which the curds are cooked varies from 30 °C (e.g., no cooking) for high-moisture cheeses (e.g., Camembert, Gorgonzola, Taleggio, Italico) to 55 °C for low-moisture cheese (e.g., Emmental, Grana Padano, and Parmigiano Reggiano).

After cooking, the curds and whey are separated by various, variety-specific techniques. The curds for most varieties are transferred to molds where further whey drainage and acidification occur. Curds that have undergone extensive syneresis in the vat (e.g., have a low moisture content) are pressed in the molds, sometimes according to a programmed increase in pressure, with the objective of fusing the curds and rendering the cheeses free from mechanical openings and reducing the moisture content further.

The curds for two families of cheese, Cheddar and *pasta filata* varieties, are subjected to special treatments prior to molding. Cheddar-type cheese undergoes a process called "cheddaring." In the traditional process, the drained curds are piled into two beds at the sides of the vat, separated by a channel for whey drainage. The beds of curd are cut into blocks, ca. 10 cm side, which are inverted every 15 min and later piled two or three blocks high. This process continues for ca. 2 h, until the pH decreases to 5.4. During cheddaring, the blocks of curd flow slightly and the cheese acquires a fibrous texture similar to that of cooked chicken breast meat. In the modern mechanized process, the drained curds are transferred to a moving belt on which the mass of curds flows slightly but much less than in the traditional process. Previously, it was believed that the flow during cheddaring was essential for the texture of Cheddar, but it is likely that the most important change during cheddaring is acidification, which dissolves the colloidal calcium phosphate; when the ratio between Ca^{2+} and protein decreases to a certain value, the texture assumes the characteristics of Cheddar cheese. Acidification produces the same effect on almost all the Italian cheeses obtained through acid-rennet coagulation. When the pH and properties of the cheddared curd are satisfactory, the blocks of curd or the beds of curd on the cheddaring belt are cut (milled) into chips (cross section, ca. 2 cm and ca. 10 cm long) and salted. The uniform distribution of salt is critical; in mechanized systems, the salt is distributed over the bed of milled curd by a boom, and mixed. The salted curds are held for ca. 30 min to allow absorption of the salt and its diffusion into the curd chips.

The salted curds are then transferred to molds and pressed. In the traditional process, cylindrical or rectangular molds, with a capacity of 20 kg, were used and the molded curds were pressed overnight at 2.5 kPa and cooled to ambient temperature. Next day, the cheeses were coated with wax to prevent loss of water and mold growth. In modern mechanized plants, the salted chips of curd are moved pneumatically to towers about 10 m tall (Wincanton towers) in which they are pressed under their own weight; the residence time in the towers is ~30 min. As the column to curd exits the tower, 20 kg block are cut off by a guillotine, wrapped in waxed paper, and placed in cardboard boxes. The packed cheeses are conveyed for ca. 20 h through a tunnel at ca. 10 °C to cool the cheese and delay the growth of NSLAB.

The manufacture of Mozzarella curd is similar to that for Cheddar up to the point at which the pH decreases to 5.4. The acidified curds are then heated in hot (80–85 °C) water to 60–65 °C, kneaded, and stretched (*pasta filata* cheeses, Mozzarella, Pizza cheese, Provolone, Kashkaval, and many varieties of Balkan cheese). It is claimed that the kneading and stretching are essential for the characteristic fibrous texture and stretchability of Mozzarella. However, it may be that the function of heating and kneading is simply to inactivate enzymes and kill bacteria and, in effect, to stabilize the characteristics of the cheese. Heating and kneading were probably introduced originally to control the microbiota of cheese curd produced from milk of poor microbiological quality. Flow diagrams for some worldwide important cheese varieties are shown in Fig. 3.3.

The last manufacturing operation is salting, which for most varieties is performed by immersing the cheeses in saturated NaCl brine and less commonly by the

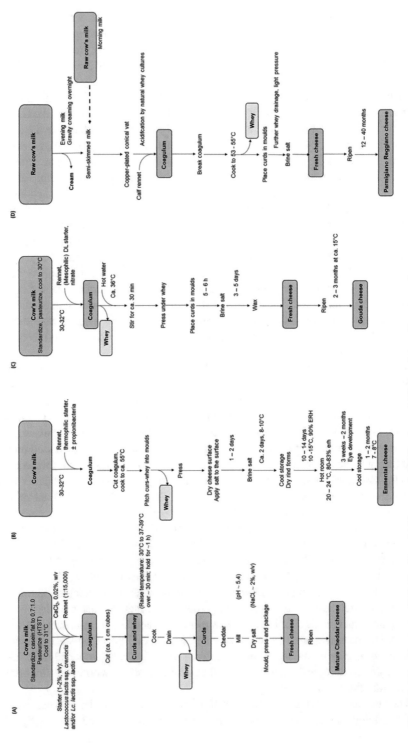

Fig. 3.3 Flow diagrams of (**a**) Cheddar, (**b**) Emmental, (**c**) Gouda, and (**d**) Parmigiano Reggiano cheeses

application of dry salt on the surface of pressed cheese or by mixing dry salt with chipped curd before pressing (used for the English varieties Cheddar and Cheshire). While salting contributes to syneresis (2 kg H_2O are lost per kg NaCl taken up), it should not be used as a means of controlling the moisture content of cheese. Salt has several functions in cheese (Guinee and Fox 2017). Although salting should be a very simple operation, quite frequently it is not performed properly, with consequent adverse effects on cheese quality.

As indicated previously, cheese manufacture is essentially a dehydration process. With the development of ultrafiltration as a concentration process, it was obvious that UF would have applications in cheese manufacture, e.g., for standardizing cheese milk with respect to fat and casein, or for the preparation of a concentrate with the composition of the finished cheese, commonly referred to as" pre-cheese." Standardization of cheese milk by adding UF concentrate (retentate) is now common and the manufacture of pre-cheese has been successful commercially for a range of soft and semisoft cheese varieties.

3.7 Ripening

Acid-coagulated cheeses are consumed fresh and such cheeses constitute a major proportion of the cheese consumed in some countries; the principal acid-curd cheeses are described in McSweeney et al. (2017). However, rennet-coagulated cheeses undergo a period of ripening (curing, maturation), which varies from a few hours/days (e.g., for Mozzarella) to 2 years (e.g., Parmigiano Reggiano or extra-mature Cheddar), the duration of ripening being generally inversely related to the moisture content of the cheese. Many varieties may be consumed at any of the several stages of maturity, depending on the flavor preferences of consumers and economic factors.

Although curds for different cheese varieties are recognizably different at the end of manufacture (mainly as a result of compositional and textural differences arising from differences in milk compositional and processing factors), the unique characteristics of the individual cheeses develop during ripening. In most cases the biochemical changes that occur during ripening, and hence the flavor, aroma, and texture of the mature cheese, are largely predetermined by the manufacturing process, e.g., by composition, especially moisture, NaCl, and pH, by the type of starter, and in many cases by secondary microorganisms added to, or gaining access to, the cheese milk or curd.

During ripening an extremely complex set of biochemical changes occur through the catalytic action of the following agents: coagulant, indigenous milk enzymes, especially plasmin and lipoprotein lipase, which are particularly important in cheese made from raw milk; starter bacteria and their enzymes; and secondary microbiota and their enzymes.

The secondary microbiota may arise from the adventitious microorganisms in milk that survive pasteurization or gain entry to the milk after pasteurization (see

Chap. 5), e.g., *Lactobacillus, Pediococcus, Micrococcus*, or they may be inoculated as a secondary starter, e.g., *Propionibacterium* in Swiss cheese, *Penicillium roqueforti* in blue cheese varieties, *Penicillium camemberti* in Camembert, or Brie, or the cheese may acquire a surface microbiota from the environment during ripening, e.g., the complex Gram-positive microflora of smear-ripened cheeses such as Tilsit, Taleggio, and Limburger. In many cases, the characteristics of the finished cheese are dominated by the metabolic activity of these microorganisms.

The primary biochemical changes which occur during ripening involve the metabolism of residual lactose and, in some varieties, of lactate and citrate, lipolysis and proteolysis which are described in Fox et al. (2017) and McSweeney et al. (2017). These primary changes are followed and overlapped by a host of secondary catabolic changes, including the various reactions involving amino acid catabolism (transamination, deamination, decarboxylation, and various lyase activities), fatty acid catabolism and related reactions (oxidation of fatty acids, esterification, formation of thioesters), and the catabolism of lactic acid to CO_2 and H_2O or to propionic, acetic, or butyric acids and CO_2 or H_2 (see also Chap. 5).

Many of the above reactions occur to some extent in all ripened cheese, but the extent and balance vary between varieties. The flavor of cheese is due to not a single compound, or even a group of compounds, but to the concentrations and balance of many compounds.

The biochemistry of cheese ripening, which has been studied extensively, has been reviewed in several chapters in Fox et al. (2017) and McSweeney et al. (2017).

References

Bansal N, Fox PF, McSweeney PLH (2007) Aggregation of rennet-altered casein cimelles at low temperatures. J Agric Food Chem 55:2125–3136

Donnelly C (2016) The Oxford companion of cheese. Oxford University Press, London

Fagan CC, O'Callaghan DJ, Mateo MJ et al (2017) The syneresis of rennet-coagulated curd. In: McSweeney PLH, Fox PF, Cotter PD et al (eds) Cheese: chemistry, physics and microbiology, 4th edn. Elsevier, Academic Press, Oxford, pp 145–177

Fox PF, Guinee TP, Cogan TM et al (2017) Fundamentals of cheese science, 2nd edn. Springer, New York

Guinee TP, Fox PF (2017) Salt in cheese: physical, chemical and biological aspects. In: McSweeney PLH, Fox PF, Cotter PD et al (eds) Cheese: chemistry, physics and microbiology, 4th edn. Elsevier, Academic Press, Oxford, pp 317–375

Harbutt J (ed) (1999) A Cook's guide to cheese. Anness Publishing Ltd, London

Harbutt J (2002) The world encyclopedia of cheese. Anness Publishing Ltd, London

Jaros D, Rohm H (2017) Rennets: applied aspects. In: McSweeney PLH, Fox PF, Cotter PD et al (eds) Cheese: chemistry, physics and microbiology, 4th edn. Elsevier, Academic Press, Oxford, pp 53–67

Mair-Waldburg H (ed) (1974) Handbook of cheese: cheeses of the world A to Z. Volkwertschaftlecher Verlag GmBH, Kempten Allgan

McSweeney PLH, Fox PF, Cotter PD et al (eds) (2017) Cheese: chemistry, physics and microbiology, 4th edn. Elsevier, Academic Press, Oxford

Mucchetti G, Neviani E (2006) Microbiologia e tecnologia lattiero-casearia. Qualità e sicurezza. Tecniche Nuove, Milan

Tunick MH (ed) (2014) The science of cheese. Oxford University Prass, Oxford

Uniacke-Lowe T, Fox PF (2017) Chymosin, pepsins and other aspartyl proteinases: structures, functions, catalytic mechanism and milk-clotting properties. In: McSweeney PLH, Fox PF, Cotter PD et al (eds) Cheese: chemistry, physics and microbiology, 4th edn. Elsevier, Academic Press, Oxford, pp 69–113

Walstra P, Wouters JF, Geurts TJ (2005) Dairy science and technology. CRC Press, Boca Raton

Waugh DF, von Hippel PH (1956) κ-Casein and the stabilization of casein micelles. J Am Chem Soc 78:4576–4582

Chapter 4
Classification of Cheese

4.1 Introduction

Starting from a limited range of raw materials (bovine, ewe, goat, or water buffalo milk, starter cultures, coagulant—rennet or acid—and salt), a large number (perhaps 1500) of cheese varieties are produced. In order to facilitate their study and to help consumers, retailers, and cheese technologists, a number of attempts have been made to classify cheese varieties into meaningful groups or families. Overall, criteria for classification include dairy species, coagulating agent (rennet or acid), texture/moisture content, matured or fresh, and microbiota (internal bacterial, surface/smear bacterial, internal or surface mold, propionic acid bacteria).

Traditional classification schemes are based principally on the rheological properties of cheeses, which are closely related to the moisture content (very hard, hard, semi-hard, semi-soft, or soft). Although this is a widely used basis for classification, it suffers from serious drawbacks since it groups together cheeses with widely different characteristics and manufacturing protocols. For instance, Cheddar, Parmigiano Reggiano, Grana Padano, and Emmental are often grouped together as hard cheeses, although they have quite different flavors and the methods for their manufacture are quite different. Attempts have been made to make this scheme more discriminating by including factors such as origin of the cheese milk, method of coagulation, cutting of the coagulum, scalding of the curds, drainage of whey, method of salting, and molding.

© Springer International Publishing AG, part of Springer Nature 2018
M. Gobbetti et al., *The cheeses of Italy: Science and Technology*,
https://doi.org/10.1007/978-3-319-89854-4_4

4.2   Schemes for Classification of Cheese

Walter and Hargrove (1972), who classified cheeses on the basis of manufacturing technique, suggested that there are only 18 distinct types of natural cheese, which they grouped into eight families under the headings, very hard, hard, semisoft, and soft.

Ottogalli (2001) classified cheeses into three main groups: *Lacticinia* (milk-like), *Formatica* (shaped), and *Miscellanea* (miscellaneous). The *Lacticinia* group includes products which are manufactured from milk, cream, whey, or buttermilk by coagulation with acid (lactic or citric), with or without a heating step. A small amount of rennet is often used to increase the firmness of the resultant coagulum (e.g., Quarg and cottage cheese). The *Lacticinia* group comprises seven families, ranging from yogurt-like products to whey-based products (e.g., Ricotta). The *Formatica* group contains most cheese varieties, all of which are coagulated by rennet. This is a large heterogeneous collection of varieties, which are divided into five classes, based essentially on moisture content: very hard, hard, semihard, semisoft, and soft, and the extent and pattern of ripening: internal—bacterial, surface white mold, internal blue mold, and surface bacterial smear. The *Miscellanea* group is a heterogeneous collection of varieties and includes processed, smoked, grated, and pickled cheeses, cheeses containing nondairy ingredients (fruit, vegetables, and spices), cheese analogues, and cheeses made using ultrafiltration technology.

Fox (1993) and Fox et al. (2000) proposed a complex classification scheme based on the following criteria:

1. Species of dairy animal: cow, sheep, goat, and water buffalo.
2. Coagulant: enzymatic (rennet), isoelectric (acid), and acid–heat.
3. Texture (moisture content): very hard, hard, semi-hard, semi-soft, and soft.
4. Characteristic ripening agent: internal-bacterial, surface mold, internal mold, and surface bacterial smear.
5. Eyes/openings: numerous large eyes, few small eyes, and irregular openings.

A modification of this scheme is shown in Fig. 4.1, indicating 15 groups. Examples of Italian cheeses are included within this classification, although some of the above 15 groups arguably do not refer to some Italian varieties. The classification of Italian cheeses is further complicated because the same variety may be commercialized after variable periods of ripening (e.g., Caciotta, Pecorino cheeses) (see also Chap. 6), thus including it in various groups of the classification reported in Fig. 4.1.

In addition, there is a range of modified cheeses and cheese-like products, including enzyme-modified cheese, dried cheeses, cheese analogues, and processed cheese. An unusual group of cheese-like products is manufactured in Scandinavia, especially Norway, by concentration of whey or a whey-milk mixture and crystallization of lactose and concentration of other solids in the whey. One could argue that such varieties are not cheeses at all but rather by-products of cheese manufacture made from whey; they are more like fudge than cheese. These products are

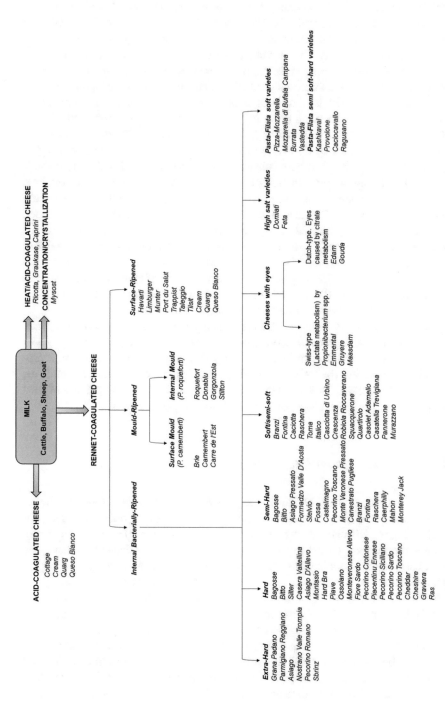

Fig. 4.1 The diversity of cheese. Cheese varieties are classified into super-families based on the principal ripening agent and/or characteristic technology

included in Fig. 4.1. These cheeses such as Brunost (brown cheese) or Mysost (Norwegian), Mesost (Swedish), Mysuostur (Icelandic), Myseost (Danish), or Braunkäse (German) are characterized by having a smooth creamy body and a sweet, caramel-like flavor. Sweet whey is the usual starting material although acid whey may be used for some varieties. Sometimes, skim milk or cream is added to the whey to give a lighter-colored product (which would otherwise be dark brown). These cheeses have a high total solids content (less than 18% moisture), are high in calories, and are characterized by a long shelf life. The whey–cream mixture is first concentrated to ca. 60% total solids (often in a multistage vacuum evaporator). A second concentration step (to >80% total solids) follows, which requires a higher vacuum. The resulting plastic mass is heated to ca. 95 °C. The Maillard reaction is encouraged during the manufacture of these cheeses and is important for the final color and flavor of the product. The concentrate is, then, cooled, kneaded, and packaged. Crystallization of lactose is controlled to avoid sandiness in the product. Several variants are produced using this basic process; differences arise from the origin of the whey (cow's milk or goat's milk), the addition to skim milk, milk or cream to the mix, or the use of sweet or acid whey.

All ripened cheeses are coagulated by rennet (ca. 75% of total world production). Acid-curd cheeses are the next most important group. Coagulation by a combination of heat and acid is used for a few varieties, including Ricotta.

Cattle, especially *Bos taurus*, are the principal dairy animals: about 85, 14, 2, and 1.5% of total milk is produced from cows, water buffalo, goats, and sheep, respectively, with small amounts of camel, yak, equine, asinine, and reindeer milk produced. About 92% of the 19 M tonnes of cheese are produced from cow's milk, ca. 3% (523,000 tonnes), 3.4% (680,000 tonnes), and 1.5% (282,000 tonnes) of total are produced from goat, sheep, or water buffalo milk, respectively.

4.2.1 Rennet-Coagulated Cheeses

There is a great diversity of rennet-coagulated cheeses and, therefore, the further classification considers the characteristic ripening agent(s) or manufacturing technology (Fig. 4.1). The most diverse family of rennet-coagulated cheeses is that containing the internal bacterially ripened varieties, which include most hard and semi-hard cheeses. Indigenous milk enzymes and residual coagulant also play important roles in the ripening of internal bacterially-ripened cheeses. This group may be sub-divided based on moisture content (extra-hard, hard, or semihard) and on whether or not the cheese has eyes. Many varieties produced on a large industrial scale are included in this group: Parmigiano Reggiano, Grana Padano (extra-hard), Cheddar, and British territorial varieties. Internal bacterially ripened cheeses, with eyes, may be subdivided based on moisture content. The subdivision includes hard varieties (e.g., Emmental) in which the numerous, large eyes are formed by CO_2 produced on fermentation of lactate by *Propionibacterium freudenreichii* subsp.

shermanii and semi-hard (e.g., Edam and Gouda), in which a few small eyes develop due to the formation of CO_2 by fermentation of citrate by a component of the starter.

Most varieties classified in groups other than internal bacterially-ripened cheeses are soft or semi-hard. *Pasta filata* cheeses (e.g., Mozzarella and variants thereof, Provolone) are very important Italian varieties, now produced worldwide. Mold-ripened cheeses are subdivided into surface mold-ripened varieties (e.g., Camembert or Brie), in which ripening is characterized by the growth of *Penicillium camemberti* on the surface, and internal mold-ripened cheeses (blue cheeses), in which *Penicillium roqueforti* grows in fissures throughout the cheese mass (e.g., Stilton, Danablue, Roquefort, and Gorgonzola). Surface smear-ripened cheeses are characterized by the development of a complex microbiota consisting initially of yeasts and ultimately of bacteria (particularly coryneforms) on the cheese surface during ripening (e.g., Limburger, Munster, Trappist, and Taleggio). White, brined cheeses, including Feta and Domiati, are ripened under brine and have a high salt content and, consequently, they are grouped together as a separate category.

The sub-division between hard and semi-hard cheeses is rather arbitrary. There is also some crossover between categories: Gruyère is classified as an internal bacterially ripened variety with a few eyes but it is also characterized by the growth of a surface microbiota, while some cheeses classified as surface-ripened (e.g., Havarti and Port du Salut) are often produced without a surface biota and therefore are, in effect, soft internal bacterially ripened varieties. *Pasta filata* and high-salt varieties are considered as separate families because of their unique technologies (stretching and ripening under brine, respectively) but they are actually ripened by the same agents as internal bacterially-ripened cheeses. However, we believe that the scheme in Fig. 4.1 is a useful basis for classification and therefore the diversity of cheeses will be discussed under these headings.

A number of cheese varieties have Protected Designation of Origin (PDO) status, which recognizes a specific heritage and provides consumers with a guarantee of authenticity. Unlike commercial trademarks, PDO denomination constitutes a collective heritage and may be used by all producers of a particular cheese in a particular geographical area. PDO cheeses are protected by the European Union under various international agreements. In addition to their geographical origin, PDO denomination also certifies that the cheese has been made using a specified (usually traditional) technology.

4.2.2 Visual Appearance of Selected Cheeses

Historically, most cheese varieties evolved to have a characteristic shape and size, probably reflecting the facilities available. In most cases, the shape and size of a cheese are largely cosmetic, but in some varieties (e.g., Emmental and Camembert), size does matter. Over the recent years, the shape of many varieties has changed, e.g.., Cheddar was traditionally produced as cylinders but is now produced as rectangular blocks, because they are easier to stack and store, and to be cut into consumer portions. The visual appearance of cheese also is characteristic, e.g., granular, with eyes or mechanical openings, with molds, etc. The experienced consumer readily recognizes the type, and perhaps even the variety, of cheese from its visual appearance.

References

Fox PF (1993) Cheese: an overview. In: Fox PF (ed) Cheese, chemistry, physics and microbiology, 2nd edn. Chapman and Hall, London, pp 1–36
Fox PF, Guinee TP, Cogan TM et al (2000) Fundamentals of cheese science. Aspen Publication, Gaithersburg
Ottogalli G (2001) Atlante dei Formaggi. Hoepli, Milan
Walter HE, Hargrove RC (1972) Cheeses of the world. Dover, New York

Chapter 5
The Distinguishing Features of Italian Cheese Manufacture

5.1 Introduction

As described in Chaps. 1 and 2, Italy has a long tradition and cultural heritage on cheeses. Based on the most widespread, official and unofficial, catalogues and atlas, Italy has documented about 400–450 varieties of cheeses (Assolatte 2016; Atlante Qualivita 2013; Mucchetti and Neviani 2006; Ottogalli 2001). This number of cheeses is certainly incomplete because of the many local varieties that escape any form of cataloguing. A country with such a large tradition and diversity has undoubtedly a cultural/technical heritage and some distinguishing features for cheese manufacture, which deserve worldwide knowledge. This chapter focuses on the main distinguishing traits that characterize the Italian cheese heritage.

Notwithstanding the influence of the chemical composition of milk from different species, the main issues are dealing with the peculiar technology interventions (e.g., raw milk processing, skimming, cooking, rennet paste) and mainly with the microbiological features. Indeed, Italy is probably the country where the use of natural starter cultures is widespread. The method for preparing natural starter cultures is characteristic, having undoubtedly repercussions on the dynamics of the cheese microbiota. Because of the very large spectrum of ripening periods and the abundance of long-ripened hard and extra-hard varieties, shaping and assembly of the lactic acid bacteria microbiota is almost unique for each of the multitude of Italian cheeses. This uniqueness emphasizes the role of both starter bacteria and milk autochthonous non-starter lactic acid bacteria (NSLAB), which have fundamental importance for the characteristics of many cheese varieties. The chapter highlights the biochemical events which occur during ripening and distinguish among the Italian varieties, and between the Italian varieties and the other worldwide cheeses.

© Springer International Publishing AG, part of Springer Nature 2018
M. Gobbetti et al., *The cheeses of Italy: Science and Technology*,
https://doi.org/10.1007/978-3-319-89854-4_5

5.2   The Main Characteristics of Italian Cheeses

Although each protocol for making cheese has a number of technological operations closely related to each other, some of them are rather common to the various categories of cheese, while others are specific and represent distinguishing features of cheese manufacture. According to this premise, this section first reports the gross chemical composition of milk, and then mainly focuses on technology traits that make typical the cheesemaking of several Italian varieties. The paragraph concludes with the protocols for preparing natural culture starters, which represent one of the main distinctive traits of the Italian cheese tradition.

5.2.1   The Gross Chemical Composition of Milk

Table 5.1 shows the approximate amount of Italian cheeses manufactured in 2016, according to the species of milk. Overall, more than 500,000 tonnes of cheese are manufactured. Cow's milk cheeses represent the most abundant production (ca. 430,000 tonnes), but the abundance of the Italian heritage is also undoubtedly enriched by the manufacture of numerous and well-known ewe, goat, and water buffalo milk cheeses. For these latter types of milk, the values indicated in Table 5.1 could be particularly low because of the many local varieties that certainly are not recorded in the annual census.

The chemical composition of the milk of a given species varies with a variety of natural factors such as, for instance, breed, lactation period and number of animal lactations, season, feeding, and geographical origin. Nevertheless, it is possible to define ranges for the chemical composition in macrocomponents. The gross chemical composition of the milk from the four major species (cow, ewe, goat, and water buffalo) used for the manufacture of the main Italian cheeses are shown in Table 5.2. Although with some variations related to breed, the overall chemical composition of the milk used for making Italian cheese varieties does not particularly differ from that used in other countries. Obviously, the differences in the chemical composition among the different species (mainly content of fat and proteins) reflect on cheese yield and on some technology interventions that are strictly tailored based on the milk used for making cheeses. It is noteworthy that for several Italian cheese varieties it is common to use a blend of milk (e.g., cow's milk and ewe's milk).

Table 5.1 Approximate amount of milk produced and cheeses manufactured in Italy in 2016 (Assolatte 2016)

Animal species	Milk quantity (tonnes)	Cheese (tonnes)
Cow	10,156,290	430,000
Sheep	529,000	41,500
Goat	21,630	2000
Water buffalo	241,000	55,000

Table 5.2 Main gross chemical composition (mg/L) of the milk from the four major species (cow, ewe, goat, and water buffalo) used for the manufacture of the main Italian cheeses (adapted from Mucchetti and Neviani 2006)

	Cow			Ewe			Goat			Water buffalo		
	m^a	$\bar{X}$	M	m	$\bar{X}$	M	m	$\bar{X}$	M	m	$\bar{X}$	M
Water		870.8			822.3			883.8			808.4	
Lactose	44.0	48.0	50.0	40.6	46.1	5.69	39.6	42.9	47.0	46.0	51.5	52.8
Ashe	6.0	7.0	8.0	9.0	10.1	10.1	5.0	7.0	8.0	7.5	8.5	9.9
Protein	28.0	33.6	38.0	49.0	56.6	64.0	27.9	31.0	42.5	36.0	45.7	57.0
Fat	33.0	40.6	47.0	51.0	64.9	87.0	29.3	35.3	45.8	73.0	85.9	94.4

^{a}m minimum value, M maximum value; and $\bar{X}$ averaged values. Minimum and maximum values refer to eventual variations depending on the animal breed

5.2.2 The Main Technology Traits

Although Chap. 6 deals in detail with the main features for processing the most traditional and popular Italian cheese varieties, a number of technology traits deserve a more detailed and separate description. These mainly concern the use of raw milk, the protocol for creaming/skimming, the curd cooking, and the use of rennet paste, which is restricted to Italian cheese varieties.

5.2.2.1 Raw Milk

The manufacture of Grana cheeses (Parmigiano Reggiano and Grana Padano) uses raw cow's milk (Neviani et al. 2013; Gatti et al. 2014). The same is true for a very long list of other varieties of cow's and ewe's milk cheeses (see Chap. 6). Upon arrival at the dairy plant, raw milk has a very complex microbiota, the composition of which largely depends on milking procedures, and time and method for collection. Because of the richness in nutrients, a pH close to neutrality and the temperature, which for most Italian varieties of cheese must be well above the refrigeration (ca. 4 °C), the milk is a very suitable substrate for a multitude of microorganisms, mainly bacteria. Among this bacterial diversity, mainly mesophilic lactic acid bacteria, coming from the house microbiota and/or milk autochthonous, have a relevant role especially in cheeses during ripening. This microbial group, mostly known as non-starter lactic acid bacteria (NSLAB), has an important and positive impact on raw milk cheeses made without a starter, using natural milk or whey cultures or using commercial/selected starters. The manufacture of raw milk cheeses without using starters (e.g., Bitto cheese) reproduces ancient dairy processes based exclusively on the selection of the milk microbiota through processing parameters. Such varieties of cheese have distinctive and appreciated taste and flavor attributes but may also show the appearance of technology and microbiology defects. Usually, the long period of ripening for these cheese varieties is the main tool to guarantee safe hygiene conditions and to avoid or limit taste and flavor defects. Although these cheeses are quantitatively a marginal area of the cheesemaking, they are undoubtedly important for animal husbandry economy in several Italian mountainous areas. Raw milk cheeses manufactured using natural milk or whey cultures represent the most traditional way of milk processing in Italy. Undoubtedly, this category of cheeses is important for the national dairy industry and because of its worldwide spread. This is the case of many PDO (Protected Denomination of Origin) cheeses, for instance Parmigiano Reggiano, Grana Padano, and Mozzarella di Bufala Campana. The shaping and assembly of the natural milk or whey culture microbiota and its interaction with NSLAB result from many and balanced technological interventions. Usually, raw milk cheeses manufactured with commercial/selected starters are from small farms unable for a number of reasons to manage with protocols for making natural milk or whey cultures. Starters have usually the capacity to drive the milk fermentation process and to inhibit the growth of undesired microorganisms. Nevertheless, in all three conditions (cheeses made without starters or with natural or commercial/selected starters) the milk autochthonous microbiota interacts and contributes to driving the biochemical events during cheese ripening. As a permanent link with the microbiological quality of milk and with cheesemaking environment, the composition of the milk autochthonous microbiota may vary markedly at the species and biotype levels.

In several cases, cooling of milk at the farm avoids the risk of growth of harmful or spoilage microorganisms. Refrigeration of raw milk, however, shifts the balance within the autochthonous milk microbiota. It may favor the multiplication of psychrophilic and/or psycrothrophic bacteria, which, in turn, cause the partial inhibition of the important NSLAB microbiota. As the Italian tradition of cheesemaking has successfully demonstrated, the combination of non-refrigerated milk and microbial starters is the most suitable option for exploiting the potential of the milk autochthonous bacterial patrimony.

Pasteurization causes profound changes in the raw milk autochthonous microbiota. If the pasteurization kills pathogenic bacteria and inhibits most of the spoilage microorganisms, it also markedly depauperates the bacterial autochthonous patrimony of raw milk. The thermization of raw milk represents a technological compromise to avoid pasteurization and refrigeration and to inhibit spoilage microorganisms. Usually, the thermization of raw milk is at a temperature ranging between 57 and 68 °C for at least 15 s. After this treatment, the milk has a positive phosphatase reaction, and the effect on the reduction of some microbial species depends on the intensity of heating.

Another technological option for processing the raw milk is the so-called prematuration, even if it has almost disappeared from the current protocols of cheesemaking. This operation somewhat predisposes milk to processing. Usually, commercial/selected lactic acid bacteria starters, not necessarily the same species and/or biotypes present in the natural or in the selected strains used as primary starters, are added to cheese milk. The temperature of pre-maturation is usually below that for the optimal growth of the microbial strains used. The main effect of pre-maturation is partial proteolysis that provides starter bacteria (subsequently inoculated) with low molecular weight peptides, which should speed up the following acidification process.

5.2.2.2   Skimming/Creaming

Skimming is a process of fat separation used to standardize and reduce the concentration of fat in raw milk for cheesemaking. When milk is not homogenized, the cream naturally rises to the surface because of the difference on specific gravity. Skimming of milk by natural creaming also causes a decrease of the number of spore-forming bacteria, which are dragged into the cream by the fat globules. Depending on the temperature and duration of creaming, an increase of the raw milk mesophilic microbiota also occurs. Skimming for Parmigiano Reggiano and Grana Padano cheeses and for a number of other varieties consists of the partial removal of fat from unhomogenized milk. Usually, it takes place overnight at ca. 20 °C or 12–15 °C (in the case of Grana Padano cheese) in special tanks, *bacinelle* (capacity, 10–50 hL or bigger, 300–500 hL, in the case of Grana Padano cheese), which contain a shallow body of milk (Gatti et al. 2014). Only the evening milk undergoes partial skimming. A slight microbial acidification occurs during creaming, and the number of spore-forming bacteria decreases. Depending on the temperature and duration of skimming, an increase of the milk autochthonous mesophilic microbiota also occurs. After this treatment, the partially skimmed milk is mixed, in a ratio of 1:1, with whole milk from the following morning milking. The fat content of the cheese milk for making Parmigiano Reggiano is ca. 2.4–2.5% (Gatti et al. 2014). Usually, the removed fat is the main ingredient for making traditional and much appreciated butter. The manufacture of Asiago and Montasio cheeses uses an almost similar protocol for skimming as that described for Parmigiano Reggiano and Grana Padano.

5.2.2.3   Breaking and Cooking the Curd

Cooking the curd has a direct effect on whey separation and drainage because it causes a large increase of the capacity of curd to synerese. Indirectly, curd cooking affects the preparation of the whey natural cultures, because of the relatively high temperature that selects the composition of the resulting autochthonous microbiota. The intensity of the heat treatment varies depending on the variety ca. 40–43 °C for Italico, ca. 46° for Asiago, ca. 48 °C for Montasio, and up to 56 °C for Grana Padano and Parmigiano Reggiano cheeses. According to this variable range of temperature, this treatment is differentiated into cooking (cooked cheeses) and low cooking (low-cooked cheeses), when the low heating is applied. In general, the temperature and the time of heating depend on the expected residual moisture content of the cheese and it is, therefore, associated with the extent of the curd rupture. From a microbiological point of view, cooking the curd drives the early stages of cheese fermentation and ripening. Indeed, a thermal selection of lactic acid bacteria, which are naturally present in the milk or added as starters, occurs and varies depending on the cooking intensity (Neviani et al. 1995). This treatment represents a classical example where the technology interferes with the microbiology features of cheesemaking of traditional Italian cheese varieties.

5.2.2.4   Rennet Paste

A special form of animal rennet is lamb paste (or simply rennet paste), which is traditional in some Mediterranean countries, and in Italy in particular, for the manufacture of mainly typical ewe's cheeses like Pecorino Romano and Fiore Sardo (Mucchetti et al. 2009). Traditionally, the manufacture of lamb paste is from milk-filled stomachs of freshly slaughtered lambs. These stomachs are then milled, dried, salted, and ripened (Addis et al. 2008). Slaughtering conditions and ripening temperature have the main influence on the enzyme activity of lamb pastes. Lamb paste contains a high amount of pre-gastric and gastric esterases that facilitate lipolysis and the formation of flavor-intense free fatty acids during cheese ripening. This intense and piquant flavor is one of the main sensory attributes distinguishing several Italian varieties of ewe's milk cheeses (mainly Pecorino cheeses) subjected to a medium or long ripening period. This sensory attribute is achievable only using rennet paste. When producing rennet paste on an artisanal level, especially in some regions for the manufacture of typical cheese varieties, the microbiological load of the preparations is an important issue. Frequently, the enzyme pattern of rennet paste, as influenced by its raw materials and production conditions, largely determines also the intensity of proteolysis and, consequently, the flavor and texture development of such cheeses. Therefore, skills on rennet paste preparation as well as its long-time use are indispensable to avoid defects and to exploit the potential of this type of renneting.

5.2.3   Milk or Whey Natural Culture Starters

The use of starters is a practice adopted in Italy since the beginning of the last century, with the main aim of reducing defects of microbiological origin in cheeses. Currently, it has consolidated over time, becoming a very common practice for almost all the industrial production.

Several schemes classify the starter cultures for cheesemaking, mainly based on function, temperature of growth, bacterial composition, and method for production. Primary starters (exclusively lactic acid bacteria) are involved mainly in the synthesis of lactic acid from lactose, which occurs early in cheese manufacture. The overall rule is to get a decrease in the pH of milk to less than 5.3, in 6 h at 30–37 °C, depending on the cheese variety. Therefore, the addition of large numbers of active cells of primary starters into cheese milk is usual. The secondary microbiota is more diverse and belongs to the main groups as follows: (1) NSLAB, consisting of non-starter lactobacilli, *Pediococcus*, *Enterococcus*, and *Leuconostoc*; (2) propionic acid bacteria; (3) molds; and (4) bacteria and yeasts, which grow on the surface of smear-ripened cheeses. Together with primary starters, the secondary microbiota mainly contributes to the cheese sensory characteristics (Lazzi et al. 2016). There is evidence that lysis of primary starters (natural and/or selected/commercial), mainly during cheese ripening, should release nutrients (e.g., peptides and free amino acids), which favor the subsequent growth of the secondary microbiota, mainly NSLAB. Because these microorganisms may have a role during ripening, their natural contamination, from milk and environment, is often enough for several cheese varieties (Sgarbi et al. 2013). Nevertheless, the need for standardization and acceleration of ripening has prompted the use of many secondary starters, often called adjuncts (mainly NSLAB), to enhance the sensory features of cheeses (Gobbetti and Di Cagno 2017; Bancalari et al. 2017).

Primary starters are usually classified as mesophilic or thermophilic. These latter are characteristics of many Italian cheese varieties, where a high temperature (>37 °C, up to 48–52 °C for the hard cooked varieties) prevails during the early phases of cheese manufacture. Mesophilic starters are used in all Italian cheese varieties where the temperature of the curd during the early stages of lactic acid production does not exceed ca. 40 °C. However, this distinction is losing part of its significance, since mesophilic and thermophilic species are sometimes found (or used) together in both undefined and defined starters for making some Italian cheese varieties (Beresford et al. 2001).

All starter cultures available today derive from natural starters of undefined composition, reproduced daily in cheese factories by some form of back slopping. This tradition is continuing for many Italian cheese varieties, representing one of the main and antique distinguishing features. The use of natural whey cultures for making Grana cheeses was introduced in Italy in a fully empirical way in 1888 (Mucchetti and Neviani 2006; Neviani et al. 1998; Bottazzi 1993). For other Italian cheese varieties, commercial defined or undefined starters replaced the natural cultures. Usually, these correspond to mixed-strain starters derived from the selection of the best performing natural starters reproduced under controlled conditions by specialized institutions (dairy research centers or commercial starter companies). Because of their optimized and highly reproducible performance, and their high phage resistance, defined-strain starters have replaced traditional starters in the production of some cheeses, including some PDO varieties (e.g., Gorgonzola, Taleggio, Pecorino Toscano, and Quartirolo cheeses).

Overall, the reproduction of traditional natural starter cultures occurs daily at the cheese plant by some sort of back slopping (e.g., the use of an old batch of a fermented product to inoculate a new one) and/or by application of selective pressure (heat treatment, incubation temperature, low pH). No special precautions prevent contamination from raw milk or the cheesemaking environment, and the control of media and culture conditions during starter reproduction is limited. As a result, in any given cheese plant, natural starters are stable or may evolve continuously, depending on the selective conditions used during preparation. The composition and techniques for the production of natural starter cultures distinguishes two subtypes, milk and whey culture starters, depending on the medium and techniques used for their reproduction (Muchhetti and Neviani 2006; Parente 2006). In Italy, the standards of identity of many PDO cheeses require the use of natural starter cultures, because a strict relationship exists between their use and cheese quality. It is estimated that 12–20×10^3 tonnes of natural starter cultures are used per year for the manufacture of PDO varieties. Approximately 50–60% of the total amount of the cheese manufactured in Italy requires the use of natural culture starters. These are a valuable source of strains with desirable technological features (synthesis of volatile compounds and antimicrobial, and phage resistance) (Carminati et al. 2011). Because no attempts are made to prevent contamination, phages are invariably present, and strains isolated from natural starters are usually lysogenic and bacteriophage infection affects population and community structure by selecting for bacteriophage resistant strains (Carminati et al. 2011). Figure 5.1 shows the flow charts for their production.

5.2.3.1   Natural Milk Cultures

In Italy, the manufacture of several traditional cheeses (e.g., Asiago, Pecorino Siciliano, Canestrato Pugliese, Castelmagno, Fossa) uses or may use (as stated in the official protocols for cheesemaking) natural milk cultures (*coltura naturale in latte*, *lattoinnesto* or *lattofermento*) (Fig. 5.1). These cultures are produced daily from raw milk, using a selective heat treatment (15–30 min) and incubation at a high temperature (60–65 °C), with back slopping. A rapid cooling follows the heat treatment. Usually, *Streptococcus thermophilus* dominates these cultures but other thermophilic species may be present such as *Streptococcus macedonicus*, *Enterococcus*, and *Lactobacillus* spp. (Andrighetto et al. 2002; Delgado et al. 2013).

5.2.3.2   Natural Whey Cultures

A large number of Italian cheese varieties use or may use (as stated in the official protocols for cheesemaking) natural whey cultures (e.g., Grana Padano, Parmigiano Reggiano, Provolone, Montasio, Pecorino Sardo, Fiore Sardo, Caciocavallo Silano, Mozzarella di Bufala Campana). Natural whey cultures (*siero-fermento* or *sieroinnesto*) are prepared by incubating whey drained from the cheese overnight under more or less selective conditions (Fig. 5.1). When cooking occurs on the curd prior to removal from the vat, heating selects the autochthonous microbiota and, consequently, the microbial vat ecosystem. Therefore, the composition of the natural whey cultures results as the sum of bacteria coming from raw milk and the lactic acid bacteria introduced in the previous batch of cheese with the natural starter. Under this traditional protocol, the whey represents the link between cheeses, which are manufactured every following day. The propagation technique and the composition of the natural whey cultures for Parmigiano Reggiano and Grana cheeses have been reviewed recently (Gatti et al. 2014; Neviani and Gatti 2013). The main drivers for the selection of lactic acid bacteria are the high incubation temperature (mostly 55 °C under a temperature gradient), the low pH at the end of incubation (below 4.0), and the back slopping due to reinoculation of milk. Contrarily to natural milk cultures, the dominance of cultures in whey is by aciduric and thermophilic strains. Usually, *Lactobacillus helveticus* represents more than 85% of the cultivable lactic acid bacteria, but other thermophilic species (*Lactobacillus delbrueckii* subsp. *lactis*, *Lactobacillus fermentum*, and *Str. thermophilus*) are present. The number of

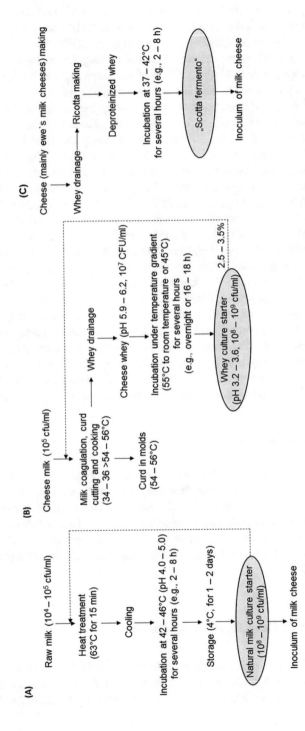

Fig. 5.1 Examples of traditional protocols for the preparation of natural milk (*coltura naturale in latte, lattoinnesto or lattofermento*) (**a**) and whey (*sierofermento or sieroinnesto*) (**b**) culture starters, and *scotta-fermento* (**c**) as used for making several Italian cheese varieties

dominant biotypes is variable (usually between two and eight) (Gatti et al. 2003; Rossetti et al. 2008; Santarelli et al. 2008; Bottari et al. 2010). Seasonal and plant-to-plant variations in the composition and performance of the culture occur. The clustering of 24 natural whey cultures used for making Grana Padano cheeses defined four typologies depending on which of these species dominated (Rossetti et al. 2008). Other studies on natural whey cultures for making Parmigiano Reggiano cheeses (Bottari et al. 2013) reported the same four species as the dominant ones. The profiles of natural whey cultures for water buffalo Mozzarella cheeses revealed the presence of *Lb. helveticus*, *Lb. delbrueckii*, *Str. thermophilus*, and *Enterococcus durans* (Ercolini et al. 2012; De Filippis et al. 2014; Silva et al. 2015). The technology of Grana and Parmigiano Reggiano cheeses and that of Provolone cheese differ and strains isolated from the cheeses were distinguished based on both phenotypic and genotypic characteristics (Gatti et al. 1999, 2004; Giraffa et al. 2000). *Str. thermophilus* strains from the whey natural starters and isolated from Asiago d'Allevo, Montasio, and Monte Veronese, having almost similar technology, were not separated by RAPD-PCR (Random Amplified Polymorphic DNA-Polymerase Chain Reaction) (Andrighetto et al. 2002). Overall, the main remark is concerning that only a fraction of the community is detectable by cultivation-dependent techniques (Bottari et al. 2013; Pogacic et al. 2013; Santarelli et al. 2013a, b; Bottari et al. 2010; Rossetti et al. 2008; Santarelli et al. 2008). A recent study based on next generation sequencing techniques has confirmed that the diversity of natural starter cultures is relatively low, that a core microbiome exists for each type of cheese, and that some species do not necessarily play a major role in cheese (De Filippis et al. 2014). However, especially for ripened cheese varieties, the low number of NSLAB found in natural starter cultures may contribute to cheese ripening (Bottari et al. 2013).

5.2.3.3   Deproteinized Whey Cultures

Other types of whey cultures include deproteinized whey cultures (*scotta-innesto*) (Fig. 5.1) used for the manufacture of Pecorino Romano cheese and other Pecorino varieties (Mannu et al. 2002). Invariably, thermophilic lactobacilli (*Lb. helveticus*, *Lb. delbrueckii* subsp. *lactis*) dominate the cultures produced under highly selective conditions (high temperature) while streptococci (*Str. thermophilus*, but also lactococci and enterococci) are found in cultures incubated at a relatively low temperature (less than 42 °C).

5.3   Dynamic of the Cheese Microbiota

The dynamic of the microbiota, which characterizes the Italian cheese varieties, has several peculiar features linked to the specific cheesemaking processes and some traits that, inevitably, are almost common with most of the other worldwide varieties.

5.3.1 General Traits

Overall, lactic acid bacteria such *Lactobacillus*, *Lactococcus*, *Enterococcus*, and *Streptococcus*, with minor populations of *Pseudomonas*, *Hafnia*, *Clostridium*, *Leuconostoc*, *Faecalibacterium*, *Prevotella*, *Acinetobacter*, and *Aeromonas* spp., dominate the microbiota of almost all raw milk species, mainly cow's milk (Quigley et al. 2013a, b). The eukaryotic inhabitants of raw milk are fungi, including yeasts and molds (Panelli et al. 2013, 2014). Milk from a healthy udder is virtually sterile and is thought to be colonized via numerous sources (e.g., teat surface, machinery, farm staff, environment, feed and collection vessels) upon milking (Tolle 1980). Other factors, stage of lactation and seasonality, may also influence the composition of the milk microbiota (Siefarth and Buettner 2014). After collection, the refrigeration of raw milk causes a decline in the growth of the majority of microbes but a boom in psychrotolerant bacteria, including *Acinetobacter* spp. and *Pseudomonas* spp. (von Neubeck et al. 2015). Pasteurization, when applied, considerably reduces psychrotrophic and mesophilic populations, though there is evidence that a proportion of these microorganisms enters a viable but noncultivable state, which results in an underestimation of their levels by culture-based methods. Spores and a variety of other thermoduric organisms survive pasteurization and gain entry into the cheese, along with other microorganisms whose introduction is via post-pasteurization contamination (Quigley et al. 2013a, b).

5.3.2 Variety-Specific Features

The peculiar manufacturing processes, which differentiate the Italian cheese varieties, play the major role in defining the final environmental conditions that somewhat counteract the factors listed in the previous paragraph. The most representative example is the curd cooking up to 57 °C for Grana cheeses, or the preparation of the natural whey cultures at a temperature above 45 °C for several hours, which inevitably affects the growth of starters and undesirable microorganisms. Overall, the primary environmental factors controlling the growth of microorganisms in cheese include water and salt content, pH, the presence of organic acids and bacteriocins, redox potential, and ripening temperature (Beresford et al. 2001).

Not differently from other cheese processing, also the microbial populations of Italian cheeses is classified in the starter populations, which drive the ripening process, and secondary microbiota, which makes each cheese type distinct (Beresford et al. 2001). Raw milk cheeses tend to have a shorter ripening time and more distinctive, varied rich flavor tones than cheeses made from pasteurized milks due to their microbial load (Montel et al. 2014). The variety of different types of raw milk cheeses manufactured makes a unique core microbiota somewhat difficult to elucidate; for example, hard and soft cheeses, short or long ripened cheeses, non-starter or starter fermented cheeses, and brined or unbrined cheeses, all have an almost distinct microbiota (Delcenserie et al. 2014; Montel et al. 2014; Wolfe et al. 2014).

An abundant literature has dealt with the dynamics and assembly of the cheese microbiota during ripening of the Italian varieties (Mucchetti and Neviani 2006). The techniques for characterization used culture-dependent and -independent approaches, with the latter that evolved rapidly regarding the capacity for describing microbial phyla, genera, and species. Consequently, some results for the same cheese variety are not perfectly comparable during time due to the different techniques used. Starter bacteria are the most prevalent microorganisms in cheese, particularly early in ripening. There is a large consensus about the dominance of thermophilic lactic acid bacteria when using natural milk or whey cultures. Primary starters need to be active in milk during curd manufacture. Both acid production and rapid growth (from ca. 10^6 cfu/mL, in the inoculated milk, to ca. 10^9 cfu/g, in the cheese curd) occur. Based on the length of ripening, the microbiota undergoes alteration, with members of the natural milk or whey cultures outcompeting the original populations from the raw milk as time proceeds. Indeed, many groups of NSLAB flourish post-brining and during ripening (Cogan et al. 2007; Gatti et al. 2014; Neviani et al. 2013). To facilitate reading, the further description combines cheeses in almost homogeneous categories.

5.3.2.1   Cooked Extra-hard Cheese Varieties

This paragraph describes the dynamics of the microbiota in extra-hard cheese varieties, which mostly share the use of raw cow's milk, partial skimming, natural starter cultures, curd cooking at different temperatures, and ripening for at least 12 months. The combination of culturing and hybridization studied the flux of *Lb. helveticus*, *Lb. delbrueckii* subsp. *lactis*, and *Lb. delbrueckii* subsp. *bulgaricus* in experimental extra-hard Grana Padano cheese and revealed that, while *Lb. helveticus* was the dominant species in the natural whey starter and during the early stages of ripening, *Lb. delbrueckii* subsp. *lactis* became dominant by month 2 of ripening. By the end of the ripening period, *Lactobacillus casei*, *Lactobacillus paracasei*, and *Lactobacillus rhamnosus* had become dominant (Zago et al. 2007). These patterns were also apparent in subsequent investigations on commercial Grana Padano cheeses (Santarelli et al. 2013a, b; Pogacic et al. 2013). The highest bacterial diversity in Grana Padano cheese occurs at 2 months of ripening. At this time, the death of thermophilic starters and the growth of NSLAB intersect. Cultivable NSLAB increased significantly in 2-, 6-, and 9-month-old cheeses, and then decreased significantly in 13-month-old cheese (Santarelli et al. 2013a, b). Culture-based investigations on Parmigiano Reggiano cheese revealed that during ripening the numbers of non-starter lactobacilli decreased from 10^8 cfu/g after 5 months to approximately 10^4 cfu/g at 24 months (Coppola et al. 1997). *Lb. paracasei*/*Lb. casei* and *Lb. rhamnosus*, which persisted throughout the 24-month ripening, dominated the population and *Pediococcus acidilactici* was present for 22 months. Pediococci appear to be fundamental for maintaining the equilibrium within the cheese-related microbial community, probably also having a negative correlation with the growth of clostridia. These findings are consistent with those derived using culture-independent approaches (Levante et al. 2017; Gatti et al. 2014; Sgabi et al. 2014). These showed that starters from natural whey cultures such as *Lb. helveticus* and *Lb. delbrueckii* subsp. *lactis* dominate until 1 month after brining, at which point *Lb. rhamnosus* or *Lb. casei* and *P. acidilactici* became detectable. The same species were present after 6 months of ripening, although none was clearly dominant, and no further changes occurred between this time and month 20 of ripening (Gatti et al. 2008, 2014). *Lb. paracasei* was dominant in Montasio cheese, and although it was not detectable immediately after manufacture, its population increased to 10^7 cfu/g during the first month of ripening and remained at that level up to 120 days (Lombardi et al. 1995).

5.3.2.2 Ewes' Pecorino Cheese Varieties

Although manufactured and consumed also as fresh varieties, most of the traditional Italian Pecorino cheese varieties use raw or pasteurized ewe's milk, alone or in combination with a low percentage cow's milk, natural starter cultures, and are subjected to a considerable period of ripening (at least 4–6 months). Together with components of the NSLAB microbiota such as *Lb. curvatus*, *Lb. plantarum*, and *Lb. fermentum*, and with a heterogeneous population of enterococci, the thermophilic *Lb. dekbrueckii* subsp. *lactis* was found occasionally in ripened Pecorino Romano (Di Cagno et al. 2002). Recent evidence showed that the raw ewe's milk used for making this cheese was contaminated by several bacterial phyla, but from day 1 of ripening onwards, the phylum *Firmicutes* dominated. After 90 days, mesophilic lactobacilli (e.g., *Lb. plantarum*) and *Lactococcus lactis* dominated and established a causal relationship with proteolysis (De Pasquale et al. 2014a). Italian PDO Fiore Sardo, Pecorino Siciliano, and Pecorino Toscano ewe's milk cheeses were the hard varieties chosen as model systems to investigate the spatial distribution of the metabolically active microbiota and the related effects on proteolysis and synthesis of volatile components (De Pasquale et al. 2016). Gradients for moisture, and concentrations of salt, fat and protein distinguished sub-blocks, while the cell density of the main microbial groups did not differ. Secondary proteolysis differed between sub-blocks of each cheese, especially regarding hydrophilic and hydrophobic peptides and free amino acids. Regardless of the cheese variety, the profile with the lowest level of volatile components was restricted to the core region of cheeses. As shown through pyrosequencing of the 16S rRNA targeting RNA, the spatial distribution of the metabolically active microbiota agreed with the distribution of volatile components. Top and bottom under rind sub-blocks of all three cheeses harbored the widest biodiversity. The cheese sub-block analysis revealed the presence of a microbiota statistically correlated with secondary proteolysis and/or synthesis of volatile components. The dynamic of the cheese microbiota of Canestrato Pugliese cheese made from ewe's milk or with a combination of ewe's milk and cow's milk, was subjected to characterization. Here, enterococcal numbers decreased by one to two orders of magnitude during the 9-week ripening period (Albenzio et al. 2001), while the non-starter *Lactobacillus* population increased to 28 days and remained constant for the remainder of maturation. The microbial population in cheese made from raw ewe's milk (10^8 cfu/g) was approximately three log cycles higher than that in cheeses made from thermized or pasteurized milk. Marked shifts in the species profiles characterized also ewe's milk Pecorino Sardo cheese (Mannu et al. 2002). Non-starter lactobacilli and enterococci increased during the 60-day ripening period, although significant differences occurred between batches. Numbers of *Lb. casei* were almost constant throughout, becoming dominant when the lactic acid bacteria starter population declined. The numbers of non-starter lactobacilli in traditional farmhouse Fiore Sardo cheese increased from 10^5 cfu/g in 1-day-old cheese to 10^8 cfu/g after ripening for 30 days. The population then decreased slowly, and by 7 months, the level decreased to 10^4 cfu/g (Mannu et al. 2000). *Lb. plantarum* decreased markedly during ripening, while *Lb. paracasei*, when present, dominated the cheese microbiota.

5.3.2.3 *Pasta filata* Cheese Varieties

The common traditional feature of these cheeses is stretching the curd under hot whey, at an elevated and variable temperature, and manufacture as fresh or long-ripened cheese. Changes occur within the microbial community of the Sicilian artisanal *pasta filata*-type cheese, Ragusano (Randazzo et al. 2002). Mesophilic lactic acid bacteria, including *Leuconostoc* spp. and *Lc. lactis*, dominated the raw milk population, but disappeared during cooking and curd ripening. *Lb. delbrueckii* and *Lb. fermentum* grew during ripening, and enterococci were present in reasonable numbers as the microbial population stabilized in 30-day-old cheeses. Major shifts in species profile also occurred during the ripening of another *pasta filata* cheese, Caciocavallo Pugliese. The two dominant species at the end of the 60-day ripening period were *Lactobacillus parabuchneri* and *Lb. paracasei*, while *Lb. fermentum* was the dominant species in the young cheese (Gobbetti et al. 2002). The proportion of *Enterococcus faecalis* and *E. durans* in the population decreased from approximately 6% to 0.1% during ripening, while that of *Pediococcus pentosaceus* increased. Caciocavallo Pugliese and Canestrato Pugliese have also been the aim of more recent studies, which focused on RNA levels as a means of identifying active populations (De Pasquale et al. 2014a, b). In both cases, a reasonably diverse microbial population in milk became less diverse as the cheesemaking process continued until eventually streptococci and lactobacilli became dominant. Indeed, the only atypical genera detected in Caciovallo Pugliese were low levels of *Paracraurococcus*, *Pseudoalteromonas*, and *Rhodococcus* (after 30 days of ripening), and *Azospirillum* and *Gelria* (after 75 days of ripening) (De Pasquale et al. 2014b). The identification of a larger number of sub-dominant populations characterized the Canestrato Pugliese cheese, with *Enterococcus* (after 90 days of ripening) and *Carnobacterium* (up to 75 days of ripening) constituting over 1% of the total population (De Pasquale et al. 2014a). Recently, the *pasta filata* Mozzarella cheese, made from raw water buffalo milk and natural whey cultures, underwent characterization through new generation sequencing techniques (De Filippis et al. 2014). *Str. thermophilus* was the most abundant species, and, while *Lb. fermentum* was detected within the starter community, it and *Lc. lactis* were found to be present at low levels only and not in all samples.

5.3.2.4 Other Cheese Varieties

Studies on the microbiota dynamic of other heterogeneous Italian cheese varieties are presented here. The microbiology of semi-hard Fontina cheese has been studied using a variety of molecular approaches. The relatively high diversity of bacteria in cheese milk became less diverse as the cheesemaking process continued with lactic acid bacteria starters ultimately dominating. Investigations also revealed the presence, at low levels, of *Pseudomonas*, *Staphylococcus*, and *Enterobacteriaceae* in most of the cheeses at 84 days of ripening (Dolci et al. 2014). A combination of PCR-DGGE (Denaturing Gradient Gel Electrophoresis) and RT (Real Time)-PCR-DGGE has also been used to investigate the microbiology of the hard variety Castelmagno cheese, from Northern Italy, manufactured with raw cow's milk. Both molecular approaches clearly demonstrated the key role for *Lc. lactis* in manufacture and throughout ripening. *Lb. helveticus* was also detectable at days 3, 60, and 150 of the ripening process, despite its absence (not detectable) in a parallel culture-based investigation. *Lb. plantarum* and *Lb. curvatus* were the species most frequently isolated from the semi-hard Fossa cheese, typically ripened in pits (Gobbetti et al. 1999). Fewer numbers of *L. paracasei* subsp. *paracasei* were detectable also. Using culture-dependent approaches, the lactic acid bacteria microbiota was characterized for the most famous Italian blue cheese, Gorgonzola (Gobbetti et al. 1997a). The 86-day-old cheese showed high numbers of *Str. thermophilus*, *P. roqueforti*, lactoccci, and surface growth of micrococci, yeasts, and molds. This composition of the microbiota caused an initial higher pH in the cheese surface; the final value was ca. 6.8 in both the surface and inner layers. The concentration of water-soluble N in the core was more than 50% of the total N but was markedly less in the external region. Using almost the same approaches, Taleggio cheese was characterized during ripening (Gobbetti et al. 1997b). Analyses of the surface, middle, and core layers of the ripened cheese revealed high numbers of thermophilic lactic acid bacteria and surface growth of yeasts and molds. A moderate increase of micrococci and contaminating bacteria also occurred at the cheese surface.

All the above differences and changes in the microbial population are relevant factors, which affect Italian cheeses during subsequent ripening, especially regarding the extensive secondary proteolysis, which leads to an elevated concentration of small peptides and amino acids, and which undoubtedly relates to the peptidase activity of mesophilic bacteria.

5.4   Non-starter Lactic Acid Bacteria (NSLAB)

All Italian varieties, especially those subjected to some extent of ripening, harbor an autochthonous microbiota, which results mainly from environmental (milk and house microbiota) and technology factors, and which mainly ensures the diversity and the typical feature, together with natural starters, between cheeses throughout processing (Montel et al. 2014; Quiegly et al. 2013a, b). Such cheese autochthonous microbiota is composed mainly of so-called NSLAB. Lactococci, pediococci, enterococci, *Leuconostoc* sp., and thermophilic lactic acid bacteria are part of this population, but NSLAB consist mainly of mesophilic facultative and obligate hetero-fermentative lactobacilli.

The origin of NSLAB is always under discussion. Certainly, raw milk is the main source of NSLAB. It provides the vat milk with its microbiota and enriches the microbiota of the cheesemaking environment (Montel et al. 2014). Although enriched with thermophilic primary starters, natural whey cultures are also sources of NSLAB (Gatti et al. 2014). NSLAB have been isolated from floors and drains in the dairy environment, and from surfaces of equipment used in cheese manufacture and vacuum packaging (Bokulich and Mills 2013). The capacity of several mesophilic lactobacilli (e.g., *Lb. plantarum*) to form biofilms enhances their survival after cleaning and disinfecting treatments, and the persistence of a specific house microbiota in the dairy environment (Somers et al. 2001). Post-pasteurization contamination has been described (Beresford et al. 2001), but some NSLAB strains also withstand heat treatments, mainly resulting in damaged cells that recover and proliferate in the curd during ripening (De Angelis and Gobbetti 2004). The milk and house microbiota should ensure an almost constant NSLAB supply. Nevertheless, cell density and biodiversity vary during subsequent cheese batches and, especially, decrease for pasteurized milk cheeses.

The diversity of mesophilic lactobacilli is quite wide and the cheesemaking performance of the lactobacilli species may differ markedly. Compared to primary starters, NSLAB have the opposite kinetic of growth in cheese. After curd manufacture, their number is typically 10^2–10^3 cfu/g, which increases slowly and reaches a plateau at 10^7–10^9 cfu/g after a few to several months of cheese ripening (Fitzsimons et al. 2001). NSLAB have a generation time of ca. 8 days during ripening at 6 °C (Jordan and Cogan 1993). When cheese aging is prolonged (e.g., Parmigiano Reggiano), the maximum cell density decreases but NSLAB autolysis has rarely been described (Gatti et al. 2014; Lazzi et al. 2016). Species and, in particular, biotypes of NSLAB may vary between dairy pants, within a dairy plant depending on season and day of manufacture, and even vary between batches of cheese. Rarely, a single NSLAB may predominate throughout the entire cheese ripening period (Coolbear et al. 2008; Levante et al. 2017; Bove et al. 2011). Figure 5.2 summarizes some information, based on an abundant but certainly not exhaustive literature (Gobbetti et al. 2015). Eighteen Italian cheese varieties and/or variants, manufactured from milk of different species, are considered. Commonly, cheeses are from raw milk but varied for the species of milk and technology treatments such as cooking, stretching, pressing, with or without the use of commercial or natural starters, and length of ripening. Overall, thirteen species of mesophilic lactobacilli were variously identified. Facultative

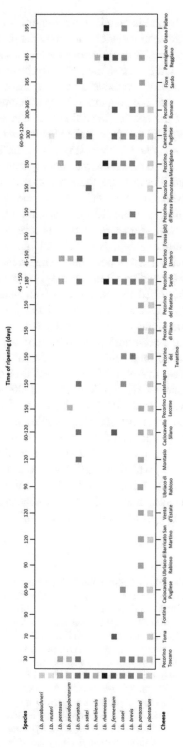

Fig. 5.2 Diversity of mesophilic lactobacilli species, mainly as members of the nonstarter lactic acid bacteria (NSLAB) population, in several Italian cheese varieties. Information regarding the variety of cheese and the age of ripening is also given. Adapted from Gobbetti et al. (2015)

heterofermentative lactobacilli dominate. *Lb. paracasei* or *Lb. plantarum* were present in seventeen out of the eighteen varieties, their association being the most common. In spite of the biotype variability, the above species constitute, together with *Lb. curvatus*, *Lb. rhamnosus* and *Lb. casei*, the core microbiota of the NSLAB population. The other species represent a variable portion of this microbiota, in most of the cases constituting a sub-dominant population. Only for some varieties (seven out of eighteen), the number of the species identified was one or two. Cooking and stretching the curd commonly and mainly correlated with the presence of *Lb. casei*, *Lb. rhamnosus,* and *Lb. parabuchneri*, supposing a specific thermal tolerance of these species. The use of natural starters correlated mainly with the appearance of *Lb. rhamnosus* and *Lb. fermentum*. For instance, these two species are frequently present as natural contaminants of the natural whey cultures for making Grana cheeses (Gatti et al. 2014). Middle to long-time ripening (120 days onwards) correlated mainly with *Lb. plantarum* and *Lb. paracasei*, which are the two most representative NSLAB.

Growing in cheese during ripening, NSLAB undergo several environmental stresses (e.g., acidity, starvation, redox potential). Their adaptation responses are efficient but vary markedly depending on the species and strains. In the literature, only a few reports have described the NSLAB as responsible for defects (e.g., abnormal flavor profile, gas slits and lactate crystal formation) in cheese (Crow et al. 2002). Comparing the manufacture of the same cheese variety from raw versus pasteurized milk or defining a relationship between the consistent appearance of NSLAB versus the biochemical and sensory properties of the mature cheeses, a very abundant literature agrees that NSLAB are essential for developing the diversity and typical features of many varieties (Coolbear et al. 2008; Montel et al. 2014; Lazzi et al. 2014, 2016; Levante et al. 2017). For instance, secondary proteolysis in Canestrato Pugliese showed the most complex profiles in cheese manufactured from raw milk. Compared to pasteurized milk cheese, that made from raw milk had the highest concentration of free amino acids and water-soluble extracts contained the highest peptidase activities, which coincided with the highest cell density of NSLAB (Albenzio et al. 2001). A recent study on the same Canestrato Pugliese cheese clearly established a causal relationship between mesophilic lactobacilli and proteolysis (De Pasquale et al. 2014a). As shown in Fig. 5.3, mesophilic lactobacilli

---

(N) and area (A) of hydrophilic (7–53) and hydrophobic (53–100) peptide peaks, concentrations of volatile components (arbitrary units of area) positively correlated with the abundance of *Lb. paracasei* during ripening of Caciocavallo Pugliese cheese. Euclidean distance and McQuitty's criterion (weighted pair group method with averages) were used for clustering. The colors correspond to normalized mean data levels from low (white) to high (blue). The color scale, in terms of units of standard deviation, is shown at the top. Cheese (C) (post brine, C1) and during ripening (3, 7, 15, 30, 45, 60, 75, and 90 days, C3 e C90). *L. Lactobacillus, Str. Streptococcus, Hexadienal* 2-4 hexadienal, *3Mebutanal* 3-methyl-butanal, *Bnzald* Benzaldehyde, *2Propanol* 2-propanol, *2Butanol* 2-butanol, *2Pentanol* 2-pentanol, *2Heptanol* 2-heptanol, *3M-3buten1ol* 3-methyl-3-buten-1-ol, *2-3C8* 2-hexanedione, *3M-2C5* 2-methyl-3-pentanone, *C5H8O* cyclo-pentanone, *4Heptanone.* 4-heptanone, *IPC2* isopropyl acetate, *MB3C2* 3-methyl-butyl acetate, *CDS* carbon disulfide, *Thiole* thiophene, *2Mefuran* 2-methyl-furan, *2Ethylfuran* 2-ethyl-furan, *2Propylfuran* 2-propyl-furan, *2Butylfuran* 2-butyl-furan, *3Mefuran* 3-methyl-furan, *2,5DMefuran* 2,5-dimethyl-furan, *Acetald* Acetaldehyde, *Diacetyl* 2,3-butanedione, and *2e3C5* 2,3-pentanedione. Data from De Pasquale et al. (2014b)

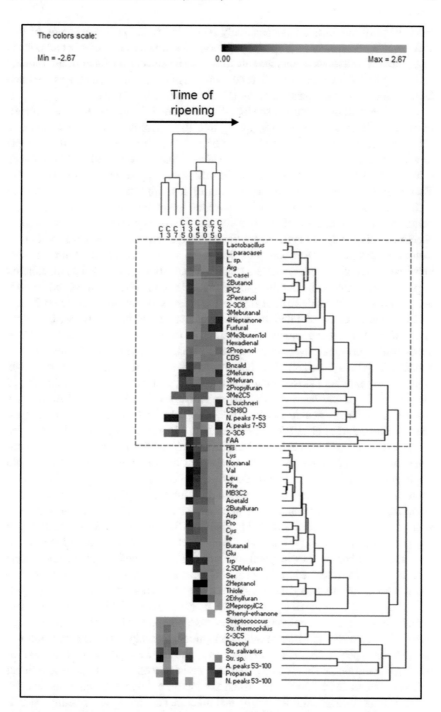

Fig. 5.3 Permutation analysis based on the relative abundance of Operational Taxonomic Units (OTUs) assigned to genus (*Lactobacillus* and *Streptococcus*) and species (*Lactobacillus paracasei, Lactobacillus casei, Lactobacillus buchneri, Streptococcus thermophilus,* and *Streptococcus salivarius*) level, total free amino acids (FAA, mg/kg) and amino acids found at the highest concentration (>100 mg/kg) (Arg, His, Lys, Val, Leu, Phe, Asp, Pro, Cys, Ile, Glu, Trp, Ser), number

were the only microorganisms positively correlated with the concentration of free amino acids, the area of hydrophilic peptide peaks, and several volatile compounds (e.g., alcohols, ketones, esters and all furans). The selection of a core microbiota, consisting of NSLAB, occurred naturally during mid-ripening, which was the main factor responsible for cheese ripening (De Pasquale et al. 2014b).

Although cheese defects caused by NSLAB seem to be extremely rare, adventitious NSLAB may introduce variability into the ripening process. For the same cheese variety this may cause fluctuations of the characteristics between dairy plants or within the same dairy plant (Antonsson et al. 2003; Levante et al. 2017), between different days of manufacture and vats on the same day (Fitzsimons et al. 1999). Therefore, the most obvious option to control adventitious NSLAB is to preempt their cheese colonization by deliberate introduction of selected NSLAB as adjuncts cultures/starters (Coolbear et al. 2008). In some cases, this option also accelerated cheese ripening. Adjunct cultures are those added to cheese for purposes other than lactic acid production (El Soda et al. 2000). This role and definition almost coincided with that of secondary starters (Chamba and Irlinguer 2004). The inoculation of pasteurized milk with combinations of mesophilic lactobacilli improved secondary proteolysis and catabolism of free amino acids, sensory and hygiene attributes, and, in some cases, accelerated the ripening of a variety of Italian cheeses (Gobbetti et al. 2015). Despite the undoubted advantages, the addition of adjunct NSLAB to cheese milk may cause over acidification of the curd due to lactose fermentation in addition to primary starters. This has repercussions on increased whey drainage, which, in turn, affects cheese yield and rheology, and the sensory acceptability of the cheese. To avoid such drawbacks, the most recent research and technology transfer have been oriented to assay the potential of attenuated adjunct cultures/ starters (Bancalari et al. 2017). These are NSLAB unable to grow and synthesize significant levels of lactic acid but can still deliver active enzymes that are responsible for secondary proteolysis and catabolism of free amino acids. This would imply good retention of the attenuated cultures in cheese curd, a significant enzyme content, and adequate lysis during attenuation treatment (Klein and Lortal 1999). Historically, the first preparation of attenuated starters was reported almost forty years ago with the aim of shortening ripening (Petterson and Sjöström 1975); a copious number of reports followed. Several methods have been used for attenuation: heating, freezing-thawing, and sonication. *Lb. casei* and *Lb. plantarum* are the species most often assayed in the form of attenuated adjunct cultures. Whatever the method used, various attenuated lactobacilli species increased proteolysis and lipolysis, shortened ripening time, improved flavor, and reduced bitterness (Klein and Lortal 1999). Some representative studies have compared the methods of attenuation or the same NSLAB in the form of an adjunct and adjunct-attenuated cultures. Selected biotypes of *Lb. paracasei*, *Lb. casei,* and *Lb. curvatus* have been used for making Caciotta-type cheese as adjunct or attenuated (sonication treatment) adjunct cultures. Attenuated adjunct cultures did not increase the production of lactic acid, while adjunct cultures acidified during manufacture and throughout ripening, affecting cheese moisture and texture. The levels of ketones, secondary alcohols, and sulfur compounds were highest in the cheese manufactured with an attenuated adjunct culture (Di Cagno et al. 2011). A similar study used *Lb. plantarum*, *Lb para-casei*, and *Lb. casei* for the manufacture of Caciocavallo Pugliese cheese. As shown by the fluorescence determination of live versus dead or damaged cells, attenuation by sonication resulted in a portion of the cells damaged and a portion of the cells being capable of growing with time. Again, the use of an adjunct culture resulted in the over acidification, which altered the gross composition of the cheese. The major differences between cheeses were the accumulation of free amino acids and the activity of several enzymes, which were highest in the cheese made with an attenuated adjunct culture. Contrarily to adjunct cultures, attenuated adjunct cultures were suitable for accelerating the ripening of Caciocavallo Pugliese cheese without modifying the main features of the traditional cheese (Di Cagno et al. 2012).

Despite the promising results, the industrial use of attenuated adjunct cultures is not as widespread as would be expected. Some reasons could be the empirical methods for attenuation, the difficulty of application at an industrial level and the extent of attenuation, which varies markedly depending on NSLAB species and biotypes. Further research is necessary to increase knowledge about these efficient cheese additives, not excluding the use of cell-free cytoplasmic extracts as reservoirs of multiple and complementary cheese ripening enzymes (Calasso et al. 2015).

5.5 Cheese Ripening

Flavor and texture development during cheese ripening has been studied intensely, including for several Italian cheeses. The function of the bacterial metabolism, the main driver for cheese ripening, has been studied, with increasing adoption of genomic and other "omic" analyses in recent years (e.g., Broome 2007; Folli et al. 2018; Gobbetti et al. 2007; Jardin et al. 2012; Liu et al. 2010; Lazzi et al. 2014; Smit et al. 2005; Steele et al. 2013). Schematics of the metabolic pathways of starter and non-starter bacteria are easily retrievable from sites, such as the KEGG pathway database (http://www.genome.jp/kegg/pathway.html) or the BioCyc database (http://biocyc.org).

5.5.1 Lactose Metabolism

As a representative example, the lactic acid fermentation of Parmigiano Reggiano cheese, during the first 48 h after manufacture, has been described comprehensively (Mora et al. 1984). There are a few comparable data for other Italian extra-hard cheeses, but the fermentation is generally similar in those varieties, which underwent cooking at a high temperature and have a rather large size. The sudden depletion of lactose also characterized the early manufacture of Castelmagno cheese (Bertolino et al. 2011). The growth of the starter thermophilic lactic acid bacteria and the hydrolysis of lactose depend mainly upon the rate at which the curds cool after removal from the cheese vat. Depending on the weight of the cheese, the temperature at the center of the curd remains relatively high, e.g., higher than 50 °C for 12–16 h for Parmigiano Reggiano, while the exterior of the cheese cools rather suddenly (ca. 2 h) to ca. 42 °C. Consequently, bacterial growth starts earlier and is more intense in the external zone. Although the depletion of residual lactose takes place throughout the cheese within 8–10 h, bacterial numbers, pH, and lactic acid concentration do not attain equal values in the center and exterior of the cheese for a much longer period. The concentration of lactic acid may also vary during ripening, attaining values in the range 0.8–1.2%.

5.5.2   *Proteolysis*

Proteolysis is the major event during ripening of most Italian cheese varieties. Essentially, the pattern of proteolysis is unique to each particular cheese variety due to differences in milk composition and manufacturing practices (particularly time/temperature profiles), which decide moisture content, pH, and the activity of proteolytic enzymes. Nevertheless, some main mechanisms are common to all varieties. Usually, the entire process distinguishes primary (hydrolysis of caseins to various sized oligopeptides) and secondary (hydrolysis of oligopeptides to short peptides and mainly free amino acids) proteolysis, and catabolism of free amino acids. Residual coagulants, natural milk proteinases, and proteinases and peptidases of starters and secondary microbiota variously contribute to change in the texture and the production of flavor and aroma compounds through the synthesis of low molecular weight peptides and amino acids, and their subsequent catabolism.

Residual chymosin activity retained in the curd is the major source of primary proteolysis in Italian cheeses made with mesophilic starters and a low cooking temperature or no cooking (e.g., Fossa, Canestrato Pugliese, Gorgonzola, Taleggio, and Caciotta cheeses). In varieties made with a high cooking temperature such as Grana or stretched *pasta filata* cheeses, chymosin undergoes more or less inactivation during manufacture, thus having a very limited role. Of the indigenous proteinases that are present in milk, the most important is plasmin, which is a heat-stable enzyme with a pH optimum of 7.5, and consequently plasmin is highly active in high-temperature cooked cheeses (e.g., Grana cheeses) or pasteurized milk cheeses, and its activity increases in curds with pH during ripening. At high cooking temperatures, activation of plasminogen also occurs, due to heat-induced inactivation of inhibitors of the plasminogen activator and probably of plasmin inhibitors (Farkye and Fox 1990). Somatic cells, recruited into milk to fight mastitic infection, comprise lysosomes that contain a number of proteinases, including cathepsins D, whose activity in cheese is rather limited.

Lactic acid bacteria are fastidious microbes that have multiple amino acid auxotrophy. Depending on the species/strain, cheese lactic acid bacteria require from 6 to 14 different amino acids for growth (Chopin 1993; Letort et al. 2002). It has been calculated that free amino acids and low molecular weight peptides, which are typically present in milk, support only limited growth (10–20% of the potential biomass) (Thomas and Pritchard 1987). Therefore, further growth requires the hydrolysis of milk proteins to amino acids through a complex proteolytic system, which fulfils their nutritional requirements and inadvertently contribute to the flavor of cheese. The main components of the proteolytic system of lactic acid bacteria are the cell envelope-associated proteinases, which depending on the cheese manufacture may variously contribute to primary proteolysis, peptide transport systems, and a range of intracellular endo- and exo-peptidases, which are almost the only sources of enzymes for secondary proteolysis. Proteolytic enzymes from other bacteria and fungi, used as starters or present as natural contaminants, may contribute also to casein and peptide breakdown during ripening depending on the cheese variety.

5.5.2.1 Primary Proteolysis

Analysis of the water-insoluble fraction of various cheeses by urea-PAGE (polyacrylamide gel electrophoresis) gives insight into the differences in peptide profiles between cheeses. Representative examples for a number of Italian cheeses described further in Chap. 6 are shown in Fig. 5.4a. Several differences are evident. In many cheeses where chymosin is not inactivated by cooking and a residual activity remains in the curd, α_{s1}-casein is hydrolyzed faster than β-casein (Sousa et al. 2001), while in Parmigiano Reggiano, β-casein is hydrolyzed faster than α_{s1}-casein, with concomitant increases in γ-caseins that indicate a role of plasmin and denaturation of chymosin during cooking. Indeed, plasmin and thermophilic *Lactobacillus* cell-envelope proteinases are responsible for extensive primary proteolysis in Parmigiano Reggiano cheese ripened for a long period (ca. 24 months) (Battistotti and Corradini 1993). Almost the same kinetic of casein degradation was observed also for Grana Padano and Pecorino Romano cheeses. One-year-old cheeses generally do not contain β-casein, whereas at the end of ripening Parmigiano Reggiano cheese still contains unhydrolyzed α_{s1}-casein. The γ-caseins–β-casein ratio has been proposed as an index of proteolysis in Parmigiano Reggiano during ripening (Addeo et al. 1988). During the first year, the γ-caseins represent ca. 20% of the oligopeptides, γ_1-casein being ca. 30% of the total γ-caseins. After this period, the percentage of γ_1-casein decreases, while that of γ_2- and γ_3-caseins increases due to the hydrolysis of the former by plasmin. SDS (Sodium dodecylsulfate)–PAGE and a specific anti-β-casein monoclonal antibody identified γ_1- and γ_2-casins in Grana Padano cheese during ripening, showing a correlation with the extent of ripening (Gaiaschi et al. 2001). Nevertheless, the same authors found that the geographical area of cheese-making, season of production, length of ripening and type of dairy are all factors which may affect proteolysis. In blue-veined cheeses such as Gorgonzola, after sporulation, enzymes from *Penicillium roqueforti* hydrolyze α_{s1}-casein (f24-199) and several other peptides (Gripon 1993), and all of both α_{s1}- and β-caseins are hydrolyzed at the end of ripening. The urea-PAGE profiles of pH 4.6-insoluble

84   5 The Distinguishing Features of Italian Cheese Manufacture

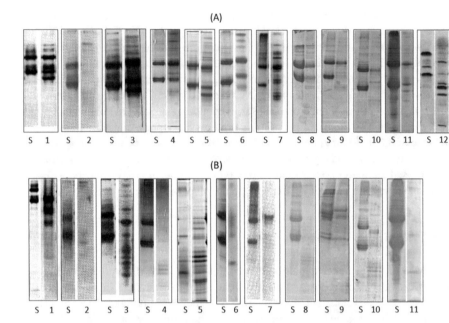

Fig. 5.4 Urea-polyacrylamide gel electrophoresis (PAGE) of pH 4.6-insoluble (**a**) and -soluble (**b**) nitrogen fractions of several Italian cheeses at different times of ripening. Lanes: S, bovine or ewe's casein standard; (*1*) Taleggio (42, after 42 days of ripening) (Gobbetti et al. 1997a); (*2*) Gorgonzola (86 days) (Gobbetti et al. 1997b); (*3*) Crescenza (14 days) (Gobbetti et al. 1998); (*4*) Pecorino Crotonoese (105 days) (De Pasquale et al. 2017); (*5*) Caciocavallo Pugliese (90 days) (De Pasquale et al. 2014b); (*6*) Canestrato Pugliese (90 days) (De Pasquale et al. 2014a, b); (*7*) Fossa (90 days) (Gobbetti et al. 1999); (*8*) Pecorino Siciliano (120 days) (De Pasquale et al. 2016); (*9*) Pecorino Toscano (120 days) (De Pasquale et al. 2016); (*10*) Fiore Sardo (120 days) (De Pasquale et al. 2016); (*11*) Caciotta (60 days) (Di Cagno et al. 2011); and (*12*) Parmigiano Reggiano (18 months) (Addeo et al. 1988)

fraction of ewe's milk Fossa cheeses varies. Nevertheless, the profiles commonly showed the complete degradation of α_{s1}-casein after 6 months of ripening while much of the β-casein persisted unhydrolyzed (Gobbetti et al. 1999). The manufacture of Fossa cheese is without cooking the curd, and chymosin activity may persist during ripening. The same applies for Canestrato Pugliese (Albenzio et al. 2001; Corbo et al. 2001) cheese. In the latter case, since the option to make the cheese from raw, thermized, or pasteurized ewe's milk, RP-FPLC (reverse-phase fast protein liquid chromatography) analysis of the water-insoluble fraction showed a more complex peptide pattern in raw milk cheese, which was positively linked to more intense proteolysis. Urea-PAGE analysis indicated extensive primary proteolysis of both β- and α_{s1}-caseins for Castelmagno cheese (Bertolino et al. 2011).

5.5.2.2 Secondary Proteolysis

The degree to which caseins and the first large peptides released therefrom are hydrolyzed is measured as the amount of water or pH 4.6-soluble N compounds in cheese, and, in general, it varies from very limited (e.g., Mozzarella cheese) to very extensive (e.g., Grana cheeses and other hard varieties, and blue cheeses such as Gorgonzola). Representative examples for a number of Italian cheeses further described in Chap. 6 are shown in Fig. 5.4b. Several differences are evident. Low molecular weight peptides formed in Parmigiano Reggiano cheese during ripening have been isolated and identified using fast atom bombardment-mass spectrometry (Addeo et al. 1992, 1994, 1995). Oligopeptides originating from regions 1-20 and 6-28 of β-casein, five phosphopeptides originating from the region 64-84 of α_{s1}-casein, three phosphopeptides from the region 1-21/24 of α_{s2}-casein and one peptide from C-terminal part of α_{s2}-casein were identified. Other authors (Sforza et al. 2012), also studied in detail the dynamic composition of peptides for Parmigiano Reggiano cheese, in part confirming the above results. Limited amounts of the first small peptides occurred already during cheesemaking, some of which originated from chymosin activity (α_{s1}-CN f1-23, f24-34/36/38)), and they disappeared within 48 h. The content of the main part of the peptides identified increased during ripening to a maximum and thereafter they decreased. After half a year, phosphopeptides from β-CN (f11-28) dominated the peptide fractions, and became smaller over time and had generally disappeared after 24 months. The same development characterized Grana Padano cheese during ripening, in which the amount of phosphopeptides decreased after 8-14 months of ripening (Ferranti et al. 1997). Hydrolysis of the caseins leads to an increased proportion of water-soluble N, which was considered as a ripening index for Parmigiano Reggiano (Panari et al. 1988) (Fig. 5.5). The increase is very fast during the first 8–10 months, after which hydrolysis proceeds very slowly. At the end of ripening, the water-soluble N is ca. 34% of the total N. Similar values (ca. 32%) characterized Grana Padano cheese (Addeo and Chianese 1990). Since the pH of many extra-hard cheeses is in the range 5.0 to 5.5, the values of water-soluble and pH 4.6-soluble N do not differ significantly. Values of pH 4.6-soluble N/total N ranging from 19 to 29% were found in Pecorino Romano cheese, which coincided approximately with those for the 12% TCA-soluble N (Guinee and Fox 1984; Guinee 1985). Since the pH 4.6-soluble N is produced principally by rennet, whereas starter and nonstarter bacterial enzymes are principally responsible for the formation of 12% TCA-soluble N, these data support the view that rennet is not very active in this cheese and that once it produces soluble peptides, bacterial peptidases hydrolyze them relatively rapidly. Contradictory results described the proteolysis in Pecorino Romano, which varied with the zones of the cheese. At the beginning of ripening, some authors found greater proteolysis in the interior of the cheese, which from 40 days onward was more extensive in the surface zone due to the inward diffusion of NaCl. Other authors (Guinee and Fox 1984; Guinee 1985) did not find differences in the level of water- and pH 4.6-soluble N at various locations in the Romano-type cheese throughout ripening. The levels of pH 4.6-soluble N are very high also in Fossa cheese, ranging from 30 to 39% of the total

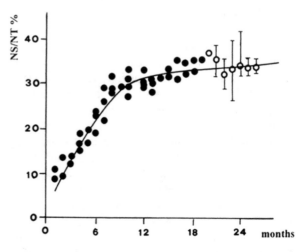

Fig. 5.5 Increase (%) in the level of water-soluble nitrogen (SN)/total nitrogen (TN) in Parmigiano Reggiano cheese during ripening. Open circles are the average of several cheeses, of the same age, at the end of ripening (Panari et al. 1988)

N. The water-soluble N may range from 13 to 30% of the total N in Canestrato Pugliese cheese, depending upon several factors, including NSLAB activity.

Overall, it is estimated that ca. 25% of the proteins in cheese are hydrolyzed into peptides and amino acids during ripening. Significant concentrations of free amino acids, the final products of secondary proteolysis, occur in all cheeses subjected to a considerable ripening period. Nevertheless, the level of free amino acids in cheese varies between varieties. The variation of the concentration of free amino acids during ripening may feature as another index (Table 5.3). Free amino acids accumulate in Parmigiano Reggiano cheese until 15 months of ripening, after which the concentration decreases at the end of ripening (Resmini et al. 1988). At the end of ripening, the average concentration of total free amino acids is ca. 230 mg/g, which corresponds to ca. 23 % of the total protein content. Therefore, Parmigiano Reggiano is one of the cheeses richest in free amino acids. A chemometric model estimated the age and the organoleptic quality of Parmigiano Reggiano, based on the level of serine, glutamine, arginine, and ornithine as markers (Resmini et al. 1988). The same trend, with similar values, characterized Grana Padano cheese, showing that the extension of ripening to more than 18 months did not produce a significant increase in free amino acids (Resmini et al. 1990). The amino acid profile of cheeses of the same age varies widely. The total concentration of free amino acids in Fossa cheese markedly varies between samples. It was estimated an average value of ca. 108 mg/g (Gobbetti et al. 1999), which is relatively high compared to Cheddar cheese that typically has ca. 3 mg/g of cheese (Lynch et al. 1996), and with internally mold-ripened cheese such as Gorgonzola, which has a value of ca. 15 mg/g (Gobbetti et al. 1998). A similar average value of ca. 104 mg/g was detectable in Canestrato Pugliese cheese manufactured from raw ewe's milk (Albenzio et al. 2001). Apart from the high concentrations of threonine, isoleucine, and phenylalanine in Parmigiano Reggiano cheese, glutamic acid, proline, valine, leucine, and lysine are the amino acids commonly present at high concentrations in Parmigiano

Table 5.3 Concentration of individual and total free amino acids (mg/g cheese) in Parmigiano Reggiano, Canestrato Pugliese, Fossa, and Fiore Sardo cheeses

Amino acids	Parmigiano Reggiano	Canestrato Pugliese	Fossa	Fiore Sardo
Histidine	8.20	3.82	2.44	0.67
Arginine	2.50	5.01	0.25	1.10
Serine	13.60	8.85	3.09	1.18
Aspartic acid + asparagine	18.60	2.99	4.09	0.46
Glutamic acid + glutamine	45.50	15.34	19.19	1.36
Threonine	12.30	3.23	2.07	0.22
Glycine	6.40	2.55	1.8	0.18
Alanine	6.90	2.87	5.83	0.53
Tyrosine	6.30	1.66	2.02	0.55
Proline	n.d.	8.65	5.6	0.08
Methionine	7.20	3.25	3.97	0.48
Valine	18.40	8.33	9.56	1.00
Phenylalanine	13.20	5.88	5.42	1.38
Isoleucine	15.90	6.54	6.24	0.59
Leucine	22.20	10.99	13.83	1.33
Cysteine	n.d.	1.57	5.00	0.06
Ornithine	3.80	n.d.	n.d.	0.47
Lysine	30.80	13.31	13.09	0.83
Tryptophan	n.d.	0.03	n.d.	0.01
Total free amino acids	231.80	104.87	103.49	12.59

The values indicated represent the average of several determinations made by different authors in cheeses, which had a slightly different ripening time: Parmigiano Reggiano, 16–18 months; Canestrato Pugliese, 6–10 months; Fossa, 6–8 months; Fiore Sardo, 4 months
n.d. not determined
Sources: Resmini et al. (1988), Gobbetti et al. (1999), Albenzio et al. (2001), Corbo et al. (2001), and De Pasquale et al. (2016)

Reggiano, Pecorino Romano, Canestrato Pugliese, and Fossa cheeses (Resmini et al. 1988; Gobbetti et al. 1999; Albenzio et al. 2001; Di Cagno et al. 2002). A decreased amount of glutamic acid and an increased content of γ-aminobutyric acid at the late stages of maturation of Fiore Sardo were concomitant phenomena. The relation between these two metabolites is due to the ability of some mesophilic lactobacilli to synthesize γ-aminobutyric acid through glutamic acid decarboxylation.

5.5.2.3   Catabolism of Free Amino Acids

Catabolism of amino acids has important implications for metabolism by starter cultures (e.g., by providing energy in the sugar-depleted environment of cheese), for the safety of cheese (e.g., by production of biogenic amines by decarboxylation of Tyr, His, Trp), and mainly for the production of flavor and aroma compounds.

Cheese flavor is the result of several non-enzymatic and many enzymatic reactions. Decarboxylation, deamination, transamination, desulfuration, and cleavage of side chains convert amino acids into aldehydes, alcohols, and acids, which together with other compounds derived by other routes (e.g., lipolysis and catabolism of fatty acids) compose the volatile profile of cheeses. In Grana Padano cheese, the catabolism of free arginine through the arginine-deiminase pathway and the synthesis of ornithine and citrulline correlated with the presence of NSLAB species (e.g., *Lb. rhamnosus*) instead of the primary starter *Lb. helveticus* (D'Incecco et al. 2016).

Based on High Resolution Gas Chromatography (HRGC)-Mass Spectrometry (MS) and different methods of extraction, volatile compounds in some Italian long-ripened cheeses have been characterized (Moio and Addeo 1998; Di Cagno et al. 2002). Overall, marked variations for the same cheese variety were found related to cheesemaking practices, season of manufacture, time of ripening, and type of secondary biota. Esters were the main neutral constituents in the aqueous distillate of Grana Padano cheese, constituting ca. 41% of the total neutral volatiles (Moio and Addeo 1998). Esters with a few carbon atoms have a perception threshold 10-fold lower than the alcohol precursors. Ethyl esters of butanoic, hexanoic, octanoic, and decanoic acids represent ca. 95% of the total esters. Ethyl hexanoate, with a distinct aroma of unripe apples, is present in the greatest quantity, ca. 60% of the total esters. This odorant in 12-month-old Grana Padano cheese is ten times the level found in fresh bovine milk (Moio et al. 1993). Ethyl butanoate is the second most important ester. Esters are the main volatile components of Canestrato Pugliese cheese (Di Cagno et al. 2002), and ethyl esters are the predominant esters in Fiore Sardo and Pecorino Romano cheeses (Moio et al. 1993; Di Cagno et al. 2002). The same was found for Castelmagno and Fossa cheeses, where ethyl hexanoate, ethyl octanoate, and ethyl decanoate dominated (Gioacchini et al. 2010; Bertolino et al. 2011). Most esters have floral and fruity notes and may positively contribute to cheese aroma. For several cheeses, ester formation correlated with the growth and enzyme activities of starter and nonstarter lactic acid bacteria.

Ketones represent the second largest class of volatile compounds in Grana Padano cheese, accounting for ca. 33% of neutral volatiles, similar to the amount found in Parmigiano Reggiano cheese where they are the most abundant volatiles, representing ca. 26% of total headspace chromatographic area (Moio and Addeo 1998; Barbieri et al. 1994). The total concentration of methyl ketones in Parmigiano Reggiano (0.075 μmol/g fat) is quite low compared to blue cheese (5.18 μmol/g fat for Roquefort) (Arnold et al. 1975; Gallois and Langlois 1990). Ketones were also the dominant volatile flavor compounds in Fiore Sardo (Di Cagno et al. 2002). Almost the same characterized the Castelmagno cheese (Bertolino et al. 2011). The major representatives of the 2-alkanones with odd numbers of carbon atoms in Grana Padano cheese were 2-pentanone, 2-heptanone, 2-nonanone, and 2-undecanone. Two-heptanone and 2-nonanone were the two methyl ketones found at the highest level in Canestrato Pugliese, Fiore Sardo, and Pecorino Romano cheeses (Di Cagno et al. 2002). All the methyl ketones with an odd number of carbons (C3-C9) were detectable in Pecorino Sardo and Fiore Sardo cheeses at higher levels than those with an even number of carbons (C4-C12) (Di Cagno et al. 2002). In Pecorino

Sardo cheese, the concentration of methyl ketones generally increases during ripening. It has been presumed that the free fatty acids liberated through lipolysis were catabolized to methyl ketones by microbial activity.

Alcohols represent the third class of volatiles in Grana Padano cheese, accounting for ca. 23% of the total neutral volatiles. Those present in greatest quantity are 2-pentanol, 3-methyl-3-buten-1-ol, 3-methyl-1-butanol, and 2-heptanol. 1-Octen-3-ol is a key aroma compound of mushrooms and is an important flavor compound produced by *Penicillium roqueforti* in blue cheeses such as Gorgonzola (Shimp and Kinsella 1977). Alcohols are the predominant group of volatile compounds in Pecorino Romano cheeses (Di Cagno et al. 2002). Parmigiano Reggiano cheese contains at least 16 different chiral alcohols, the most abundant secondary alcohols found being 2-butanol, 2-pentanol, 2-heptanol, 2-nonanol, and 1-octen-3-ol (Mariaca et al. 2001).

Aldehydes and lactones contribute ca. 0.6 and 0.1%, respectively, of the total neutral volatiles of Grana Padano cheese (Moio and Addeo 1998). A low level of aldehydes indicates a normal maturation; at higher levels, they cause off-flavor. Lactones are the second largest class of volatiles in several Italian ewe's milk cheeses, such as Canestrato Pugliese, Fiore Sardo, and Pecorino Romano, δ-dodecalactone and δ-dodecanolactone being found at the highest levels (Di Cagno et al. 2002). Eleven lactones were detectable in the Parmigiano Reggiano cheese; δ-decalactone and δ-dodecalactone were the most common (Mariaca et al. 2001).

5.5.3 Lipolysis

A low level of lipolysis accompanies the ripening of most cheeses, but extensive lipolysis occurs in Pecorino Romano and other related varieties, and blue cheeses such as Gorgonzola. The length of ripening strongly influences lipolysis and since ripening changes markedly within the same variety, cheeses ready for the market may differ greatly. Lipolysis may be due to the action of the indigenous lipase in cheese made from raw milk, to the action of microbial lipases, even though the lactic acid bacteria in starter cultures have only weak lipolytic activity, or to the action of the lipases present in rennet paste used for cheesemaking of certain varieties.

Several extra-hard Italian cheeses are probably unique, in that rennet paste is commonly used. The desirable flavor, which characterizes most of the famous Pecorino cheeses (Romano, Siciliano, and Sardo) and Fiore Sardo, but not only, is due mainly to the action of pregastric esterase in rennet paste, which is used as the source of both coagulant and lipolytic agent in cheese manufacture. Rennet pastes are prepared by grinding the engorged stomachs, including curdled milk, of young calves, kid goats, or lambs, which are slaughtered immediately after suckling or pail-feeding. Usually, the stomachs and contents are held for ca. 60 days prior to grinding. The secretion of the pregastric esterase, the physiological role of which is to aid in the digestion of fat by the young animals, which have limited pancreatic lipase activity, occurs during suckling and takes place in the stomach with ingested milk. The strong, balanced piquant flavor, which characterizes Pecorino cheeses and Fiore Sardo is due primarily to the relatively high levels of short-chain free fatty acids, especially butanoic, hexanoic, and octanoic acids. Although there are some interspecies differences, lamb, calf, and kid pregastric esterases preferentially hydrolyze fatty acids esterified at the sn-3 position of glycerol (Woo and Lindsay 1984), which explains the relatively high rate of release of butanoic acid from milk fat, in which 90% of the butanoic acid is at the sn-3 position. Calf pregastric esterase does not hydrolyze mono-butyrin, but hydrolyzes di-butyrin very slowly compared to tri-butyrin (Richardson and Nelson 1967). The moderate accumulation of short-chain free fatty acids characterizes the ripening of Parmigiano Reggiano, Canestrato Pugliese, and Fossa cheeses, for which rennet paste is not used (Woo and Lindsay 1984; Carboni et al. 1988; Gobbetti et al. 1999; Albenzio et al. 2001).

The most abundant acids found in mature Castelmagno cheese are acetic, decanoic, dodecanoic, hexanoic and octanoic acids. Acetic acid could have a microbial origin as a product of lactose fermentation. The others are derived through the activity of endogenous milk esterases and lipases (Bertolino et al. 2011). Table 5.4 shows the free fatty acid profile of some Italian extrahard cheeses. The average values refer to ripened cheeses with a high popularity on the market, but in general, no standard flavor for such extra-hard Italian cheeses is acceptable to all segments of the population. For Pecorino Romano cheese, there is a direct relationship between flavor intensity and the concentration of butanoic acid (Richardson and Nelson 1967), but the relationship between flavor desirability and butanoic acid concentration is more variable. The relative proportions of the various free fatty acids mainly influences flavor desirability. A strong, balanced, piquant Pecorino Romano cheese may be characterized by ca. 10,500 mg/kg of total FFAs, principally butanoic (C4:0), together with hexanoic (C6:0), tetradecanoic (C14:0), hexadecanoic (C16:0), and octadecenoic (C18:1) acids (Table 5.4). Among these compounds, butanoic and hexanoic acids are the most important components of the aroma quality of Pecorino Romano cheese. The total free fatty acid content of Parmigiano Reggiano approaches 20% of that generally found in Pecorino cheeses, with variations in the proportions of free fatty acids. Congeners of C18 fatty acids dominate the free fatty acids profile at the end of ripening (Carboni et al. 1988). A crude vacuum distillate of Grana Padano cheese contains large amounts of butanoic and hexanoic acids, which represent 50 and 35% of the total free fatty acids, respectively. These two free fatty acids may be important for the background aroma of Grana Padano cheese. A small change in the relative proportions of butanoic and hexanoic acids took place between 12 and 24 months of ripening (Moio and Addeo 1998). Canestrato Pugliese and Fossa cheeses show very similar free fatty acid profiles, although the former has a higher total concentration of free fatty acids (Gobbetti et al. 1999; Albenzio et al. 2001). Butanoic acid, which occurs at the highest concentration, hexanoic, decanoic (C10:0), hexadecanoic and congeners of C18 acids dominate. Probably due to the lipolytic activity of molds which colonize the cheese surface during the early period of ripening, Canestrato Pugliese also shows a rather high proportion of octadecenoic and octadecadienoic (C18:2) acids. A qualitative and semi-quantitative comparison of the free fatty acid profiles of other extra-hard varieties produced from

Table 5.4 Concentration of individual and total free fatty acids (mg/kg cheese) in Parmigiano Reggiano, Pecorino Romano, Canestrato Pugliese, and Fossa cheeses

Fatty acid	Parmigiano Reggiano	Pecorino Romano	Canestrato Pugliese	Fossa
Butanoic (C4:0)	172	3043	425	247
Hexanoic (C6:0)	48	1428	178	123
Octanoic (C8:0)	44	429	42	55
Decanoic (C10:0)	107	1009	98	84
Dodecanoic (C12:0)	107	690	46	35
Tetradecanoic (C14:0)	225	778	85	62
Hexadecanoic (C16:0)	565	1306	172	137
C18 congeners	1033	1843	322	251
Total free fatty acids	2301	10,526	1368	994

C18 congeners refer to octadecanoic (C18:0), octadecenoic (C18:1), octadecadienoic (C18:2), and octadecaatrienoic (C18:3) acids
The values indicated represent the average of several determinations made by different authors in cheeses, which had a slightly different ripening time: Parmigiano Reggiano, 16–18 months; Pecorino Romano, 10–12 months; Canestrato Pugliese, 6–10 months; Fossa, 6–8 months
Sources: Woo and Lindsay (1984), Carboni et al. (1988), Gobbetti et al. (1999), and Albenzio et al. (2001)

ewe's milk showed that butanoic, hexanoic, octanoic (C8:0), and decanoic acids were the dominant free fatty acids in Pecorino Sardo and Fiore Sardo; levels were highest in the last cheese (Larráyoz et al. 2001; Di Cagno et al. 2002).

Extra-hard cheeses produced without the use of rennet paste may vary greatly in the concentration of free fatty acids depending upon whether raw or pasteurized milk is used. Several studies have shown a higher level of free fatty acids in cheese made from raw milk than in cheese made from pasteurized or thermized milk. Such differences are due mainly to heat-induced changes of the indigenous lipoprotein lipase of milk and to the lipase and esterase activities of the milk microbiota, especially NSLAB, and become greater as the time of ripening increases. Studies on NSLAB (Gobbetti et al. 1996, 1997c) showed that *Lb. plantarum* contains lipase and esterase, which show a substrate specificity comparable to pregastric esterase and pancreatic lipase, and since there is a very large population of NSLAB in cheese during ripening, they may contribute to lipolysis.

References

Addeo F, Chianese L (1990) Cinetica di degradazione delle frazioni caseiniche nel formaggio Grana Padano. In: Grana Padano un Formaggio di Qualità: Studi e Ricerche Progetto di Qualità, Consorzio per la Tutela del Formaggio Grana Padano (ed), Reggio Emilia, pp 97–130
Addeo F, Moio L, Stingo C (1988) Caratteri tipici della proteolisi nel formaggio Parmigiano Reggiano. Composizione della frazione caseinica. In: Atti Giornata di Studio, Consorzio del Formaggio Parmigiano Reggiano (ed), Reggio Emilia, pp 21–40
Addeo F, Chianese L, Salzano A et al (1992) Characterization of the 12% trichloroacetic acid-soluble oligopeptides of Parmigiano-Reggiano cheese. J Dairy Res 59:401–411

Addeo F, Chianese L, Sacchi R et al (1994) Characterization of the oligopeptides of Parmigiano-Reggiano cheese soluble in 120g trichloroacetic acid/1. J Dairy Res 61:365–374

Addeo F, Garro G, Intorcina N et al (1995) Gel electrophoresis and immunoblotting for the detection of casein proteolysis in cheese. J Dairy Res 62:297–309

Addis M, Piredda G, Pirisi A (2008) The use of lamb rennet paste in traditional sheep milk cheese production. Small Rumin Res 79:2–10

Albenzio M, Corbo MR, Shekeel-Ur-Rehman S et al (2001) Microbiological and biochemical characteristics of Canestrato Pugliese cheese made from raw milk, pasteurised milk or by heating the curd in hot whey. Int J Food Microbiol 6:35–48

Andrighetto C, Borney F, Barmaz A et al (2002) Genetic diversity of *Streptococcus thermophilus* strains isolated from traditional cheeses. Int Dairy J 12:141–144

Antonsson M, Molin G, Ardö Y (2003) *Lactobacillus* strains isolated from Danbo cheese as adjunct cultures in a cheese model system. Int J Food Microbiol 85:159–169

Arnold RG, Shahani KM, Dwivedi BK (1975) Application of lipolytic enzymes to flavor development in dairy products. J Dairy Sci 58:1127–1143

Assolatte (2016) Italian dairy industry—report. Assolatte, Milan

Atlante Qualivita (2013) Food and wine. I prodotti agroalimentari e vitivinicoli italiani DOP, IGP, STG—Biologico. Qualivita Rome

Bancalari E, Savo Sardaro ML, Levante A et al (2017) An integrated strategy to discover *Lactobacillus casei* group strains for their potential use as aromatic starters. Food Res Int 100:682–690

Barbieri G, Bolzoni L, Careri M et al (1994) Study of the volatile fraction of Parmesan cheese. J Agric Food Chem 42:1170–1176

Battistotti B, Corradini C (1993) Italian cheese. In: Fox PF (ed) Cheese: chemistry, physics and microbiology, vol 2, 2nd edn. Chapman and Hall, London, pp 221–243

Beresford TP, Fitzsimons NA, Brennan NL et al (2001) Recent advances in cheese microbiology. Int Dairy J 11:259–274

Bertolino M, Dolci P, Giordano M et al (2011) Evolution of chemico-physical characteristics during manufacture and ripening of Castelmagno PDO cheese in wintertime. Food Chem 129:1001–1011

Bokulich NA, Mills DA (2013) Facility-specific "House" microbiome drives microbial landscapes of artisan cheesemaking plants. Appl Environ Microbiol 79:5214–5223

Bottari B, Santarelli M, Neviani E et al (2010) Natural whey starter for Parmigiano Reggiano: culture-independent approach. J Appl Microbiol 108:1676–1684

Bottari B, Agrimonti C, Gatti M et al (2013) Development of a multiplex real time PCR to detect thermophilic lactic acid bacteria in natural whey starters. Int J Food Microbiol 160:290–297

Bottazzi V (1993) Microbiologia lattiero-casearia. Edagricole, Bologna

Bove CG, De Dea Lindner J, Lazzi C et al (2011) Evaluation of genetic polymorphism among *Lactobacillus rhamnosus* non-starter Parmigiano Reggiano cheese strains. Int J Food Microbiol 144:569–572

Broome MC (2007) Starter culture development for improved cheese flavor. In: Weimer BC (ed) Improving the flavour of cheese. CRC/Woodhead, Boca Raton, pp 157–176

Calasso M, Mancini L, Di Cagno R et al (2015) Microbial cell-free extracts as sources of enzyme activities to be used for enhancement flavor development of ewe milk cheese. J Dairy Sci (9):5874–5889

Carboni MF, Zannoni M, Lercker G (1988) Lipolisi del grasso del Parmigiano Reggiano. In: Atti Giornata di Studio, Consorzio del Formaggio Parmigiano Reggiano (ed), Reggio Emilia, pp 113–121

Carminati D, Zago M, Giraffa G (2011) Ecological aspects of phage contamination in natural whey and milk starters. In: Quiberoni AL, Reinheimer JA (eds) Bacteriophages in dairy processing. Nova Science Publishers, Hauppage, pp 79–97

Chamba JF, Irlinguer F (2004) Secondary and adjunct cultures. Fox PF, McSweeney PHL Cogan TM, Guinee TP (eds) Cheese: chemistry, physics and microbiology. Elsevier, London

Chopin A (1993) Organization and regulation of genes for amino acid biosynthesis in lactic acid bacteria. FEMS Microbiol Rev 12:21–38

Cogan TM, Beresford TP, Steele J et al (2007) Invited review: advances in starter cultures and cultured foods. J Dairy Sci 90:4005–4021

Coolbear T, Crow V, Harnett J et al (2008) Developments in cheese microbiology in New Zealand e use of starter and nonstarter lactic acid bacteria and their enzymes in determining flavour. Int Dairy J 18:705–713

Coppola R, Nanni M, Iorizzo M et al (1997) Survey of lactic acid bacteria isolated during the advanced stages of the ripening of Parmigiano Reggiano cheese. J Dairy Res 64:305–310

Corbo MR, Albenzio M, De Angelis M et al (2001) Microbiological and biochemical properties of Canestrato Pugliese hard cheese supplemented with bifidobacteria. J Dairy Sci 84:551–560

Crow V, Curry B, Christison M et al (2002) Raw milk flora and NSLAB as adjuncts. Aust J Dairy Technol 57:99–105

De Angelis M, Gobbetti M (2004) Environmental stress responses in *Lactobacillus*: a review. Proteomics 4:106–122

De Filippis F, La Storia A, Stellato G et al (2014) A selected core microbiome drives the early stages of three popular Italian cheese manufactures. PLoS One 9:e89680

De Pasquale I, Calasso M, Mancini L et al (2014a) Causal relationship between microbial ecology dynamics and proteolysis during manufacture and ripening of Protected Designation of Origin (PDO) cheese Canestrato Pugliese. Appl Environ Microbiol 80:4085–4094

De Pasquale I, Di Cagno R, Buchin S et al (2014b) Microbial ecology dynamics reveal a succession in the core microbiota that is involved in the ripening of pasta-filata Caciocavallo Pugliese cheese. Appl Environ Microbiol 80:6243–6255

De Pasquale I, Di Cagno R, Buchin S et al (2016) Spatial distribution of the metabolically active microbiota within Italian PDO ewes' milk cheeses. PLoS One 11(4):e0153213. https://doi.org/10.1371/journal

De Pasquale I, Di Cagno R, Buchin S et al (in press) Effect of selected autochthonous nonstarter lactic acid bacteria as adjunct cuktures for making Pecorino Crotonese cheese. Int J Food Microbiol

Delcenserie V, Taminiau B, Delhalle L et al (2014) Microbiota characterization of a Belgian protected designation of origin cheese, Herve cheese, using metagenomic analysis. J Dairy Sci 97:6046–6056

Delgado S, Rachid CT, Fernández E et al (2013) Diversity of thermophilic bacteria in raw, pasteurized and selectively-cultured milk, as assessed by culturing, PCR-DGGE and pyrosequencing. Food Microbiol 36:103–111

Di Cagno R, De Pasquale I, De Angelis M et al (2011) Manufacture of Italian Caciotta-type cheeses with adjuncts and attenuated adjuncts of selected non-starter lactobacilli. Int Dairy J 21:254–260

Di Cagno R, De Pasquale I, De Angelis M et al (2012) Accelerated ripening of Caciocavallo Pugliese cheese with attenuated adjuncts of selected nonstarter lactobacilli. J Dairy Sci 95:4784–4795

Di Cagno R, Banks J, Sheehan L et al (2002) Comparison of the microbiological, compositional, biochemical, volatile profile and sensory characteristics of three Italian PDO ewes' milk cheeses. Int Dairy J 13:961–972

D'Incecco P, Gatti M, Hogenboom JA et al (2016) Lysozyme affects the microbial catabolism of free arginine in raw-milk hard cheeses. Food Microbiol 57:16–22

Dolci P, De Filippis F, La Storia A et al (2014) rRNA-based monitoring of the microbiota involved in Fontina PDO cheese production in relation to different stages of cow lactation. Int J Food Microbiol 185:127–135

El Soda M, Madkor SA, Tong PS (2000) Adjunct cultures: recent developments and potential significance to the cheese industry. Int J Dairy Technol 83:609–619

Ercolini D, De Filippis F, La Storia A et al (2012) "Remake" by high-throughput sequencing of the microbiota involved in the production of water buffalo Mozzarella cheese. Appl Environ Microbiol 78:8142–8145

Farkye NY, Fox PF (1990) Observations on plasmin activity in cheese. J Dairy Res 57:412–418

Ferranti P, Barone F, Chianese L et al (1997) Phosphopeptides from Grana Padano cheese: nature, origin, and changes. J Dairy Res 64:601–615

Fitzsimons NA, Cogan TM, Condon S et al (1999) Phenotypic and genotypic characterisation of non-starter lactic acid bacteria in mature Cheddar cheese. Appl Environ Microbiol 65:3418–3426

Fitzsimons NA, Cogan TM, Condon S et al (2001) Spatial and temporal distribution of non-starter lactic acid bacteria in Cheddar cheese. J Appl Microbiol 90:600–608

Folli C, Levante A, Percudani R et al (2018) Toward the identification of a type i toxin-antitoxin system in the plasmid DNA of dairy *Lactobacillus rhamnosus*. Sci Rep 7(1):12051. https://doi.org/10.1038/s41598-017-12218-5

Gaiaschi A, Beretta B, Ponesi C et al (2001) Proteolysis of β-casein as a marker of Grana Padano cheese ripening. J Dairy Sci 84:60–65

Gallois A, Langlois D (1990) New results in the volatile odorous compounds of French blue cheeses. Lait 70:89–106

Gatti M, Contarini G, Neviani E (1999) Effectiveness of chemometric techniques in discrimination of *Lactobacillus helveticus* biotypes from natural dairy starter cultures on the basis of phenotypic characteristics. Appl Environ Microbiol 65:1450–1454

Gatti M, Lazzi C, Rossetti L et al (2003) Biodiversity in *Lactobacillus helveticus* strains present in natural whey starter used for Parmigiano Reggiano cheese. J Appl Microbiol 95:463–470

Gatti M, Trivisano C, Fabrizi E et al (2004) Biodiversity among *Lactobacillus helveticus* strains isolated from different natural whey starter cultures as revealed by classification trees. Appl Environ Microbiol 70:182–190

Gatti M, De Dea Lindner J, De Lorentiis A et al (2008) Dynamics of whole and lysed bacterial cells during Parmigiano-Reggiano cheese production and ripening. Appl Environ Microbiol 74:6161–6167

Gatti M, Bottari B, Lazzi C et al (2014) Invited review: microbial evolution in raw-milk, long-ripened cheeses produced using undefined natural whey starters. J Dairy Sci 97:573591

Gioacchini AM, De Santi M, Guescini M et al (2010) Characterization of the volatile organic compounds of Italian 'Fossa' cheese by solid-phase microextraction gas chromatography/mass spectrometry. Rapid Commun Mass Spectrom 24:3405–3412

Giraffa G, Gatti M, Rossetti L et al (2000) Molecular diversity within *Lactobacillus helveticus* as revealed by genotypic characterization. Appl Environ Microbiol 66:1259–1265

Gobbetti M, Di Cagno R (2017) Extra-hard varieties. In: PLH MS, Fox PF, Cotter PD, Everett DW (eds) Cheese chemistry, physics & microbiology, vol 2, 4th edn. Academic, London, pp 809–828

Gobbetti M, Fox PF, Smacchi E et al (1996) Purification and characterization of a lipase from *Lactobacillus plantarum* 2739. J Food Biochem 220:227–246

Gobbetti M, Burzigotti R, Smacchi E et al (1997a) Microbiology and biochemistry of Gorgonzola cheese during ripening. Int Dairy J 7:519–529

Gobbetti M, Lowney S, Smacchi E et al (1997b) Microbiology and biochemistry of Taleggio cheese during ripening. Int Dairy J 7:509–517

Gobbetti M, Fox PF, Stepaniak L (1997c) Isolation and characterization of a tributyrin esterase from *Lactobacillus plantarum* 2739. J Dairy Sci 80:1110–1111

Gobbetti M, Corsetti A, Smacchi E et al (1998) Production of Crescenza cheese by incorporation of bifidobacteria. J Dairy Sci 81:37–47

Gobbetti M, Folkerstema B, Fox PF et al (1999) Microbiology and biochemistry of Fossa (Pit) cheese. Int Dairy J 9:763–773

Gobbetti M, Morea M, Baruzzi F et al (2002) Microbiological, compositional, biochemical and textural characterisation of Caciocavallo Pugliese cheese during ripening. Int Dairy J 12:511–523

Gobbetti M, De Angelis M, Di Cagno R et al (2007) The relative contributions of starter cultures and non-starter bacteria to the flavor of cheese. In: Weimer BC (ed) Improving the flavour of cheese. CRC/Woodhead, Cambridge, p 121156

Gobbetti M, De Angelis M, Di Cagno R et al (2015) Pros and cons for using non-starter lactic acid bacteria (NSLAB) as secondary/adjunct starters for cheese ripening. Trends Food Sci Technol 45:167–178

Gripon JC (1993) Mould-ripened cheeses. In: Fox PF (ed) Cheese: chemistry, physics and microbiology, vol 2, 2nd edn. Chapman and Hall, London, pp 111–136

Guinee TP (1985) Studies on the movements of sodium chloride and water in cheese and the effects thereof on cheese ripening. PhD thesis, National University of Ireland, Cork

Guinee TP, Fox PF (1984) Studies on Romano-type cheese, general proteolysis. Irish J Food Sci Technol 8:105–114

Jardin J, Mollé D, Piot M et al (2012) Quantitative proteomic analysis of bacterial enzymes released in cheese during ripening. Int J Food Microbiol 155:19–28

Jordan KN, Cogan TM (1993) Identification and growth of non-starter lactic acid bacteria in Irish Cheddar cheese. Irish J Agric Food Res 32:47–55

Klein N, Lortal S (1999) Attenuated starters: an efficient means to influence cheese ripening e a review. Int Dairy J 9:751–762

Larráyoz P, Addis M, Gauch R et al (2001) Comparison of dynamic headspace and simultaneous distillation extraction techniques used for the analysis of the volatile components in three European PDO ewes' milk cheeses. Int Dairy J 11:911–926

Lazzi C, Turroni S, Mancini A et al (2014) Transcriptomic clues to understand the growth of *Lactobacillus rhamnosus* in cheese. BMC Microbiol 14(1):Art. no. 28. https://doi.org/10.1186/1471-2180-14-28

Lazzi C, Povolo M, Locci F et al (2016) Can the development and autolysis of lactic acid bacteria influence the cheese volatile fraction? The case of Grana Padano. Int J Food Microbiol 233:20–28

Letort C, Nardi M, Garault P et al (2002) Casein utilization by *Streptococcus thermophilus* results in a diauxic growth in milk. Appl Environ Microbiol 68:3162–3171

Levante A, De Filippis F, La Storia A et al (2017) Metabolic gene-targeted monitoring of non-starter lactic acid bacteria during cheese ripening. Int J Food Microbiol 257:276–284

Liu M, Bayjanov JR, Renckens B et al (2010) The proteolytic system of lactic acid bacteria revised: a genomic comparison. BMC Genomics 11:36–51

Lombardi A, Cattelan A, Martina A et al (1995) Preliminary data on microbiological characterisation of Montasio and Asiago cheeses. In: Improvement of the quality and microbiology of the production of raw milk cheeses, Proceedings of second COST95/BIOTA Spain/European Commission, Brussels, pp 149–159

Lynch CM, McSweeney PLH, Fox PF et al (1996) Manufacture of Cheddar cheese with and without adjunct lactobacilli under controlled microbiological conditions. Int Dairy J 6:851–867

Mannu L, Comunian R, Scintu MF (2000) Mesophilic lactobacilli in Fiore Sardo cheese: PCR-identification and evolution during cheese ripening. Int Dairy J 10:383–389

Mannu L, Riu G, Comunian R et al (2002) A preliminary study of lactic acid bacteria in whey starter culture and industrial Pecorino Sardo ewes' milk cheese: PCR identification and evolution during cheese ripening. Int Dairy J 12:17–26

Mariaca RG, Imhof MI, Bosset JO (2001) Occurrence of volatile chiral compounds in dairy products, especially cheese—a review. Eur Food Res Technol 212:253–261

Moio L, Addeo F (1998) Grana Padano cheese aroma. J Dairy Res 65:317–333

Moio L, Dekimpe J, Etievant PX et al (1993) Neutral volatile compounds in the raw milks from different species. J Dairy Res 60:199–213

Montel MC, Buchin S, Mallet A et al (2014) Traditional cheeses: rich and diverse microbiota with associated benefits. Int J Food Microbiol 177:136–154

Mora R, Nanni M, Panari G (1984) Physical, microbiological and chemical changes in Parmigiano Reggiano cheese during the first 48 hours. Scienza e Tecnica Lattiero-Casearia 35:20–32

Mucchetti G, Neviani E (2006) Microbiologia e tecnologia lattiero-casearia. Qualità e sicurezza. Tecniche Nuove, Milan

Mucchetti G, Ghiglietti R, Locci F et al (2009) Technological, microbiological and chemical characteristics of Pannerone, a traditional Italian raw milk cheese. Dairy Sci Technol 89:419–436

von Neubeck M, Baur C, Krewinkel M et al (2015) Biodiversity of refrigerated raw milk microbiota and their enzymatic spoilage potential. Int J Food Microbiol 211:57–65

Neviani E, Gatti M (2013) Microbial evolution in raw-milk, long-ripened cheeses: Grana Padano and Parmigiano Reggiano case. In: Randazzo CL, Caggia C, Neviani E (eds) Cheese ripening: quality, safety and health aspects. Nova Science Publishers, Hauppauge, pp 133–148

Neviani E, Divizia R, Abbiati E et al (1995) Acidification activity of thermophilic lactobacilli under the temperature gradient of Grana cheese making. J Dairy Sci 78:1248–1252

Neviani E, Gatti M, Mucchetti G et al (1998) Considerazioni sul sieroinnesto naturale per Grana. Latte 23:76–85

Neviani E, Bottari B, Lazzi C et al (2013) New developments in the study of the microbiota of raw-milk, long-ripened cheeses by molecular methods: the case of Grana Padano and Parmigiano Reggiano. Front Microbiol 4:36–48

Ottogalli G (ed) (2001) Atlante dei formaggi. Hoepli, Milan

Panari G, Mongardi M, Nanni M (1988) Determinazione con metodi chimici delle frazioni azotate del formaggio Parmigiano Reggiano. In: Atti Giornata di Studio, Consorzio del Formaggio Parmigiano Reggiano (ed), Reggio Emilia, pp 85–96

Panelli S, Brambati E, Bonacina C et al (2013) Diversity of fungal flora in raw milk from the Italian Alps in relation to pasture altitude. Springerplus 2:405–415

Panelli S, Brambati E, Bonacina C et al (2014) Updating on the fungal composition in Sardinian sheep's milk by culture-independent methods. J Dairy Res 81:233–237

Parente E (2006) Diversity and dynamics of microbial communities in natural and mixed starter cultures. Aust J Dairy Technol 61:44–51

Petterson HE, Sjöström G (1975) Accelerated cheese ripening: a method for increasing the number of lactic starter bacteria in cheese without detrimental effect to the cheese-making process, and its effect on the cheese ripening. J Dairy Res 42:313–326

Pogačić T, Mancini A, Santarelli M et al (2013) Diversity and dynamic of lactic acid bacteria strains during aging of along ripened hard cheese produced from raw milk and undefined natural starter. Food Microbiol 36:207–215

Quigley L, McCarthy R, O'Sullivan O et al (2013a) The microbial content of raw and pasteurized cow milk as determined by molecular approaches. J Dairy Sci 96:4928–4937

Quigley L, O'Sullivan O, Stanton C et al (2013b) The complex microbiota of raw milk. FEMS Microbiol Rev 37:664–699

Randazzo CL, Torriani S, Akkermans ADL et al (2002) Diversity, dynamics and activity of bacterial communities during production of an artisanal Sicilian cheese as evaluated by 16S rRNA analysis. Appl Environ Microbiol 68:1882–1892

Resmini P, Pellegrino L, Hogenboom J et al (1988) Gli aminoacidi liberi nel formaggio Parmigiano Reggiano stagionato. In: Atti Giornata di Studio, Consorzio del Formaggio Parmigiano Reggiano (ed), Reggio Emilia, pp 41–58

Resmini P, Hogenboom J, Pellegrino L et al (1990) Evoluzione del contenuto quali-quantitativo di aminoacidi liberi nel formaggio Grana Padano. In: Grana Padano un Formaggio di Qualità: Studi e Ricerche Progetto di Qualità. Consorzio per la Tutela del Formaggio Grana Padano, Italy, pp 193–213

Richardson GH, Nelson JH (1967) Assay and characterization of pregastric esterase. J Dairy Res 50:1061–1065

Rossetti L, Fornasari ME, Gatti M et al (2008) Grana Padano cheese whey starters: microbial composition and strain distribution. Int J Food Microbiol 127:168171

Santarelli M, Gatti M, Lazzi C et al (2008) Whey starter for Grana Padano cheese: effect of technological parameters on viability and composition of the microbial community. J Dairy Sci 91:883–891

Santarelli M, Bottari B, Lazzi C et al (2013a) Survey on the community and dynamics of lactic acid bacteria in Grana Padano cheese. Syst Appl Microbiol 36:593–600

Santarelli M, Bottari B, Malacarne M et al (2013b) Variability of lactic acid production, chemical and microbiological characteristics in 24-hour Parmigiano Reggiano cheese. Dairy Sci Technol 93:605–621

Sforza S, Cavatorta V, Lambertini F et al (2012) Cheese petidomics: a detailed study on the evolution of the oligopeptide fraction in Parmigiano-Reggiano cheese from curd to 24 months of aging. J Dairy Sci 95:3514–3526

Sgarbi E, Lazzi C, Tabanelli G (2013) Nonstarter lactic acid bacteria volatilomes produced using cheese components. J Dairy Sci 96:4223–4234

Sgarbi E, Bottari B, Gatti M et al (2014) Investigation of the ability of dairy nonstarter lactic acid bacteria to grow using cell lysates of other lactic acid bacteria as the exclusive source of nutrients. Int J Dairy Technol 67:342–347

Shimp JL, Kinsella JE (1977) Lipids of *Penicillium roqueforti*. Influence of culture temperature and age on unsaturated fatty acids. J Agric Food Chem 25:793–799

Siefarth C, Buettner A (2014) The aroma of goat milk: seasonal effects and changes through heat treatment. J Agric Food Chem 62:1180511817

Silva LF, Casella T, Gomes ES et al (2015) Diversity of lactic acid bacteria isolated from Brazilian water buffalo Mozzarella cheese. J Food Sci 80:411–417

Smit G, Smit BA, Engels WJM (2005) Flavour formation by lactic acid bacteria and biochemical flavor profiling of cheese products. FEMS Microbiol Rev 29:591–610

Somers EB, Johnson ME, Wong ACL (2001) Biofilm formation and contamination of cheese by nonstarter lactic acid bacteria in the dairy environment. J Dairy Sci 84:1926–1936

Sousa MJ, Ardö Y, McSweeney PLH (2001) Advances in the study of proteolysis during cheese ripening. Int Dairy J 11:327–345

Steele J, Broadbent J, Kok J (2013) Perspectives on the contribution of lactic acid bacteria to cheese flavor development. Curr Opin Biotechnol 24:135–141

Thomas TD, Pritchard GG (1987) Proteolytic enzymes of dairy starter cultures. FEMS Microbiol Rev 46:245–268

Tolle A (1980) The microflora of the udder. Bulletin 120, International Dairy Federation, Brussels

Wolfe BE, Button JE, Santarelli M et al (2014) Cheese rind communities provide tractable systems for in situ and in vitro studies of microbial diversity. Cell 158:422–433

Woo AH, Lindsay RC (1984) Concentration of major free fatty acids and flavour development in Italian cheese varieties. J Dairy Sci 67:960–968

Zago M, Fornasari ME, Rossetti L et al (2007) Population dynamics in Grana Cheese. Ann Microbiol 57:349–353

Chapter 6
The Most Traditional and Popular Italian Cheeses

6.1 Introduction

More than 80% of 13.65 million tonnes of milk produced in Italy in 2016 was processed into cheeses, 52.6% of milk was used for the manufacture of cheeses with a Protected Designation of Origin (PDO), according to CE Regulation 1151/2012 (EU 2012). Table 6.1 shows the production of the principal Italian cheeses in 2016. Data for other individual cheese varieties are not official or not available due to the fragmentation of their production. In 2016, almost 390.000 tonnes of Italian cheeses were exported, representing an economic value of ca. 2,500,000,000 Euro (72% to the European Union, 13% America, 7% Asia, 5% non-UE European country, 2% Oceania, and 1% Africa) (Assolatte 2016).

PDO cheeses represent ca. 45% of total Italian cheese production. This is atypical and very high compared to nearly all other European countries, and it arises from the very large production of Grana Padano and Parmigiano Reggiano (ca. 28% of the whole dairy production). In Italy, there are 52 Italian PDO cheeses (Assolatte 2016; DOOR EU-database 2017; Mipaaf 2017; Mucchetti and Neviani 2006), but six of these cheeses represent more than 90% of total Italian PDO production: Grana Padano (35%), Parmigiano Reggiano (27%), Gorgonzola (10.5%), Mozzarella di Bufala Campana (8%), Pecorino Romano (7%), and Asiago (4%) (Table 6.1). Despite the abundance of cheese varieties, which reflects the heritage of the Italian dairy culture, only some of them are recognized and widely exported worldwide (Assolatte 2016).

This chapter focuses on the main relevant technological information which distinguishes the most traditional and popular cheeses in Italy. Such information comes directly from the protocols for cheesemaking as reported by the legal association, mostly organized as consortia or association, or from literature data. Although not exhaustive of all the varieties manufactured in Italy, the following list of cheeses includes those most relevant in terms of economic value, traditional and technological features, and cultural heritage. The listed cheeses are grouped based mainly on

© Springer International Publishing AG, part of Springer Nature 2018
M. Gobbetti et al., *The cheeses of Italy: Science and Technology*, https://doi.org/10.1007/978-3-319-89854-4_6

Table 6.1 Production of the principal Italian cheese varieties in 2016 (Assolatte 2016)

Cheese variety	Animal species	Milk quantity tonnes	Cheese production tonnes
Grana Padano	Cow	2,617,202	185,873
Parmigiano Reggiano	Cow	1,840,000	139,685
Mozzarella	Cow	na*	162,820
Pizza Mozzarella	Cow	na*	105,950
Gorgonzola	Cow	471,740	54,974
Mozzarella Bufala Campana	Water Buffalo	185,669	44,207
Pecorino Romano	Sheep	214,700	36,015
Asiago	Cow	194,150	21,070
Taleggio	Cow	66,861	8892
Montasio	Cow	56,132	5970
Provolone Val Padana	Cow	51,900	5293
Fontina	Cow	39,222	3718
Pecorino Toscano	Sheep	20,739	3652
Quartirolo Lombardo	Cow	22,692	3358
Piave	Cow	22,622	2064
Pecorino Sardo	Sheep	9279	1600

na* data not available

the species (cow, water buffalo, ewe, and goat) used for milk for cheesemaking, but also on the specific technology (*pasta filata* cheeses) as well on thermocoagulation of milk, cream, or whey used for manufacturing some dairy products. Table 6.2 reports the gross chemical composition of those cheese varieties for which data are available (Mucchetti et al. 2017; AFIDOP-IGP 2016; Manzi et al. 2007; Mucchetti and Neviani 2006).

6.2 Cow's Milk Cheeses

Cow's milk cheeses represent the major part of Italian production. In 2016, ca. 93% of the milk processed into cheeses was from cows. Cheesemaking is spread throughout Italy. Cow's milk cheeses range from extra-hard varieties to soft and unripened varieties (Mucchetti and Neviani 2006; Fox et al. 2017). This distinction is further complicated since in several cases a particular cheese may be consumed as an extra-hard variety but also after a shorter period of ripening, when the cheese is soft (see also Chap. 4). Usually, cow's milk is the only milk used for making cheese but for several varieties, the addition of ewe's or goat's milk is allowed. Excluding those varieties that have a very short or no period of ripening, the main extra-hard, hard, and semi-hard varieties share some common features (see also Chap. 5). These include: (1) use of raw milk; (2) cooking the curd, with low to high scald; (3) use of natural milk or whey cultures or commercial/selected starters, mainly consisting of thermophilic lactic acid bacteria; (4) fundamental role of milk autochthonous

Table 6.2 Gross chemical composition of the principal Italian cheese varieties

Cheese	PDO	Moisture (%)	Total protein ($N \times 6.38$) (%)	Fat (%)	NaCl (%)
Grana Padano	X	31 ± 1	33 ± 2	28 ± 2	1.5 ± 0.5
Parmigiano Reggiano	X	31 ± 1	33 ± 2	28 ± 2	1.5 ± 0.5
Asiago	X				
✓ D'Allevo		35 ± 4	28 ± 4	31 ± 4	2.5 ± 1.0
✓ Pressato		40 ± 4	24 ± 4	30 ± 4	1.5 ± 1.0
Bagoss		32 ± 7	35 ± 4	26 ± 9	3.0 ± 1.5
Bitto	X	37 ± 2	27 ± 3	34 ± 3	Not determined
Bra PDO	X				
✓ Tenero		44 ± 4	23 ± 1	29 ± 4	2.0 ± 0.5
✓ Duro		40 ± 1	29 ± 1	25 ± 3	1.5 ± 0.5
Branzi		34 ± 4	Not determined	34 ± 3	Not determined
Caciocavallo Silano	X	42 ± 3	27 ± 1	27 ± 1	2.0 ± 0.5
Canestrato Pugliese	X	36 ± 3	27 ± 3	31 ± 3	Not determined
Casatella Trevigiana	X	53–60	15–20	18–25	Not determined
Casciotta di Urbino	X	30–42	20–24	31–35	1.0 ± 0.5
Casolet dell'Adamello		52 ± 5	23 ± 2	20 ± 5	1.0 ± 0.5
Casolet		45 ± 2	25 ± 1	25 ± 2	1.5 ± 0.5
Castelmagno	X	35 ± 4	26 ± 2	33 ± 3	Not determined
Crescenza		58 ± 3	15 ± 1	21 ± 1	1.0 ± 0.5
Fiore Sardo	X	27 ± 5	30 ± 3	32 ± 5	3.0 ± 2.0
Fontina	X	39 ± 3	25 ± 1	32 ± 3	2.0 ± 0.5
Formaggelle di Valle Trompia		45 ± 5	23 ± 2	27 ± 4	1.5 ± 0.5
Formaggio di Fossa	X	34 ± 4	27 ± 2	35 ± 5	3.0 ± 1.0
Formai de Mut dell'Alta Val Brembana	X	37 ± 4	23 ± 2	33 ± 3	Not determined
Gorgonzola	X	49 ± 3	19 ± 1	28 ± 2	1.5 ± 1.0
Italico		42 ± 2	23 ± 1	30 ± 1	1 ± 0.5
Montasio	X	32 ± 3	26 ± 1	34 ± 1	2 ± 1.0
Monte Veronese d'allevo	X	31 ± 5	31 ± 5	32 ± 3	1.0 ± 0.5
Monte Veronese	X	43 ± 3	23 ± 3	30 ± 3	1.5 ± 1.0
Mozzarella di Bufala Campana	X	62 ± 3	16 ± 1	20 ± 2	1.5 ± 0.5
Murazzano	X	52 ± 4	20 ± 3	25 ± 2	Not determined
Nostrano Valtrompia	X	32 ± 4	38 ± 3	23 ± 6	2.0 ± 1.0
Ossolano	X	40 ± 1	27 ± 2	30 ± 4	1.5 ± 1.0

(continued)

Table 6.2 (continued)

Cheese	PDO	Moisture (%)	Total protein ($N \times 6.38$) (%)	Fat (%)	NaCl (%)
Pannerone		50 ± 3	25 ± 3	21 ± 3	Not determined
Pecorino Romano	X	32 ± 1	25 ± 1	33 ± 1	5.0 ± 1.0
Pecorino Sardo	X	29 ± 4	29 ± 1	36 ± 3	2.0 ± 1.0
Pecorino Siciliano	X	30 ± 4	32 ± 3	28 ± 3	4.0 ± 2.0
Pecorino Toscano	X	31 ± 6	27 ± 3	36 ± 3	2.0 ± 0.5
Piave	X	30 ± 2	29 ± 1	36 ± 2	2.0 ± 0.5
Provolone Valpadana	X	37 ± 3	27 ± 2	29 ± 2	3.0 ± 1.0
Quartirolo Lombardo	X	50 ± 1	19 ± 1	25 ± 1	2.0 ± 0.5
Ragusano	X	33 ± 5	29 ± 3	30 ± 3	4.0 ± 1.0
Raschera	X	45 ± 4	28 ± 2	23 ± 2	1.5 ± 0.5
Robiola Bresciana					
✓ Dura		49 ± 1	19 ± 1	26 ± 1	Not determined
✓ Morbida		52 ± 6	17 ± 1	27 ± 6	Not determined
Robiola di Roccaverano	X	50 ± 8	19 ± 3	25 ± 2	1.5 ± 1.0
Salva Cremasco	X	38 ± 1	25 ± 1	33 ± 1	1.5 ± 1.0
Silter	X	28 ± 3	33 ± 3	31 ± 1	Not determined
Stelvio	X	36 ± 2	24 ± 1	30 ± 1	1.5 ± 1.0
Taleggio	X	52 ± 3	19 ± 1	26 ± 1	1.5 ± 1.0
Toma Piemontese	X	46 ± 4	25 ± 1	26 ± 1	1.0 ± 0.5
Toma Piemontese semigrasso	X	50 ± 4	23 ± 1	19 ± 1	1.0 ± 0.5
Valle d'Aosta Fromadzo	X	46 ± 5	27 ± 2	19 ± 1	3.5 ± 1.5
Valtellina Casera	X	36 ± 2	34 ± 1	28 ± 1	Not determined
Ricotta from cow cheese		76 ± 4	9 ± 3	10 ± 6	Not determined
Ricotta from ewe cheese		70 ± 7	8 ± 1	18 ± 9	Not determined
Ricotta from water buffalo cheese		65 ± 5	10 ± 3	18 ± 3	Not determined
Mascarpone		43 ± 1	4 ± 1	47 ± 3	Not determined

microbiota; and (5) extent of chemical breakdown during ripening (Mucchetti et al. 2017; Gobbetti and Di Cagno 2017; Randazzo et al. 2013; Mucchetti and Neviani 2006).

6.2.1   The Grana Cheeses

Firstly, the term Grana cheese defines a cheese with a grainy structure, which has been produced in the Po Valley since the thirteenth century. Both Parmigiano Reggiano and Grana Padano are Grana cheeses. Besides these varieties, also Grana Bagozzo and Grana Lodigiano are Grana cheeses but with a very limited production. Currently, Bagozzo has practically disappeared from the market.

Grana Padano and Parmigiano Reggiano are two artisanal, traditional, and long-ripened hard cooked cheese varieties produced in Northern Italy (www.parmigiano-reggiano.it; www.granapadano.com; Mipaaf 2017; Gobbetti and Di Cagno 2017; AFIDOP-IGP 2016; Mucchetti and Neviani 2006; Addeo et al. 1997a; Neviani and Carini 1994). They are manufactured only in restricted geographical areas, delimited by official regulations and defended by the associations of cheesemakers organized in the respective Consortia. Grana Padano and Parmigiano Reggiano cheeses are made from partially skimmed raw milk through a lactic acid fermentation and subjected to slow and long ripening. The lactic acid microbiota, from raw milk and natural whey culture starters, plays a fundamental role in the achievement of the typical sensory characteristics of these cheeses (Gatti et al. 2014; Neviani and Gatti 2013). Other Italian extra-hard cheeses, having an almost Grana structure and similar technology to Parmigiano Reggiano and Grana Padano, are produced using different technologies than those permitted by the Parmigiano Reggiano and Grana Padano PDO regulation. Nevertheless, these differences cause changes in the milk and cheese microbiota, and, consequently, in part modifies the main biochemical events during cheese ripening (Gatti et al. 2014; Neviani and Gatti 2013; Mucchetti and Neviani 2006).

Ancient cheeses such as Grana Padano and Parmigiano Reggiano owe their success to the distinguishing sensory and texture characteristics, appreciated worldwide. The optimal manufacture of these artisanal varieties relies mainly on the quality of raw milk and on the optimum protocol for cheesemaking. Many cheesemaking factors influence starter and non-starter lactic acid bacteria (NSLAB), in particular, their interactions, growth kinetics, and biochemical activities, which determine the high-quality standards of these two cheeses. For instance, modifications of some processing parameters (e.g., cooking temperature and curd-handling techniques) have an effect on the selection, growth, survival, and lysis of the cheese microbiota under the hostile and stressful conditions of the cheese during manufacture and ripening.

Fig. 6.1 Parmigiano Reggiano cheese. The photo has been kindly supplied by AFIDOP Italy

6.2.1.1 Parmigiano Reggiano

The manufacture of PDO Parmigiano Reggiano cheese is according to a traditional and well-established technology in a restricted area of the Po Valley, which hosts about 4000 farms. In particular, manufacture takes place in the provinces of Parma, Reggio Emilia, Modena, and Bologna, on the left side of the river Reno, in the Emilia–Romagna region, and in the province of Mantova, on the right side of the river Po, in the Lombardy region (Figs. 6.1, 6.2, 6.3, 6.4, 6.5, 6.6, 6.7, 6.8, 6.9, 6.10, 6.11, 6.12, 6.13, 6.14, 6.15, 6.16, 6.17, 6.18, and 6.19).

For the manufacture of Parmigiano Reggiano, the cows are fed on only locally grown forage, which are carefully regulated. The ratio between forage and other feeds must be ≥1 to limit the dry matter (DM) derived from feeds, which are rich in starch and proteins. The ≥25% of the DM of the forage used must be produced on the same farm where the cheese is manufactured; ≥75% of the DM of the forage used must be produced within the district where Parmigiano Reggiano is legally produced; and ≤25% of the DM of the forage used may be produced in territories adjacent to the district. The feeding of silage as fodder is not allowed, to minimize the number of spore-forming, gas-producing bacteria in the milk; also, the storage of silage on the same farm is prohibited. For cheesemaking, the use of additives, other than rennet and NaCl, is prohibited (www.parmigiano-reggiano.it; Mipaaf 2017; Gobbetti and Di Cagno 2017; AFIDOP-IGP 2016; Mucchetti and Neviani 2006; Mucchetti et al. 1998; Neviani and Carini 1994; Parisi 1966).

After milking, the milk must be delivered to the dairy within 2 h. The milk is not refrigerated and should be maintained at a temperature not lower than 18 °C. A mixture of milk from two consecutive milkings is used. The evening milk is partially skimmed after overnight creaming at ca. 20 °C in special tanks, *bacinelle* (capacity, 10–50 hL), which contain a shallow body of milk (Fig. 6.3). A slight

Fig. 6.2 Geographical area for the manufacture of Parmigiano Reggiano cheese

microbial acidification occurs during creaming. This may favor rennet activity. At the same time, a slight extent of proteolysis produces short peptides that may favor the further growth of the natural whey culture. After creaming, the partially skimmed milk is mixed in a ratio of 1:1 with whole milk from the following morning milking (Fig. 6.4). The fat content of the milk for Parmigiano Reggiano is ca. 2.4–2.5%. The natural whey culture used as starter for Parmigiano Reggiano is prepared from whey from the previous cheesemaking, which is held under a temperature gradient (from ca. 53–55 °C to ca. 35 °C) for 24 h. The microbial composition of the natural starter is very complex, subject to environmental factors and dominated by thermophilic lactic acid bacteria (ca. 10^9 cfu/mL) such as *Lactobacillus helveticus*, *Lactobacillus delbrueckii* subsp. *lactis*, *Lactobacillus delbrueckii* subsp. *bulgaricus*, and *Lactobacillus fermentum* (Gatti et al. 2003, 2014; Giraffa et al. 1998; Mucchetti et al. 1998; Neviani and Carini 1994). The ratio of obligately homofermentative to

Fig. 6.3 Parmigiano Reggiano cheesemaking. Cheese milk skimming by natural creaming. The photo has been kindly supplied by the Consorzio Parmigiano Reggiano PDO, Italy

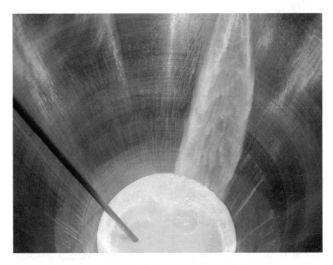

Fig. 6.4 Parmigiano Reggiano cheesemaking. Cheese milk placed in special vat (inverted bell form). The photo has been kindly supplied by the Consorzio Parmigiano Reggiano PDO

heterofermentative species is ca. 10:1 or higher. A large amount of the natural whey culture (ca. 3% v/v) is added to the milk, giving a total titratable acidity of ca. 28–32°SH/50 mL. The use of commercial/selected starters is not allowed. The calf rennet (powder preparation) used for Parmigiano Reggiano contains less than 3–4% pepsin, based on clotting activity. The time for coagulation varies from 8 to 2 min. After coagulation, the curd is broken into particles having the size of rice grains and cooked at 53–56 °C for 5–15 min under stirring. The time from rennet addition at

Fig. 6.5 Parmigiano Reggiano cheesemaking. Curd breaking. The photo has been kindly supplied by the Consorzio Parmigiano Reggiano PDO

32–34 °C to the end of cooking is 22–23 min. The combination of heating and acidifying activity by the natural whey culture allows the formation of curd grains of the right texture and whey drainage (Figs. 6.5 and 6.6). After cooking, curd grains settle to the bottom of the vat, where they congeal in ca. 30–50 min with a whey temperature not exceeding 53–55 °C. Then, the curd is removed and cut into two portions (Figs. 6.7, 6.8, 6.9, 6.10, and 6.11), each of which is placed into a mold to give the final shape (Figs. 6.12, 6.13, and 6.14). After 2 days of molding, the cheese is salted by immersion in saturated salt brine (18–22°Bé) for ca. 3 weeks (Figs. 6.15 and 6.16). The vats used for the manufacture of Parmigiano Reggiano cheese have a capacity of 10–12 hL and, traditionally, have the shape of an inverted bell (Fig. 6.4).

Fig. 6.6 Parmigiano Reggiano cheesemaking. Curd breaking. The photo has been kindly supplied by the Consorzio Parmigiano Reggiano PDO

The vats are made of copper or have an internal copper coating. The shape of the vat optimizes heat transfer, curd extraction, and molding. From each vat, two cheeses, each weighing 35–37 kg after ripening, are produced. Parmigiano Reggiano is ripened for 12 to 20–24 months at ca. 18 °C and an environmental humidity of ca. 85% (Figs. 6.17, 6.18, and 6.19).

Parmigiano Reggiano has a cylindrical shape with a diameter of 35–45 cm and height of 20–26 cm. One form of Parmigiano Reggiano has a minimum weight of 30 kg (Figs. 6.1 and 6.19). The cheese has a very low moisture content (ca. 30%), a typical compact texture, with or without many very small eyes. Cheese paste is slightly straw-colored, shows a minute grain structure, which has a radial flake fracture when cut. Aroma and flavor of the paste are characteristic: fragrant, delicate, tasty but not spicy. Aroma and flavor are the result of very slow ripening during which proteolysis is the main biochemical event (Addeo et al. 1997a; Sforza et al. 2004; Gatti et al. 2008a, b, c; De Dea Lindner et al. 2008; Gatti et al. 2014; Neviani and Gatti 2013; Manzi et al. 2007; Malacarne et al. 2006; Pecorari et al. 2007). The higher the fat content, the more the paste tends to become elastic. The rind is hard and smooth with a natural straw color. The thickness of the cheese rind is ca. 6 mm.

The cheeses on the market are the following types: Parmigiano Reggiano DOP Mezzano (ripened 12–15 months), Parmigiano Reggiano DOP (ripened 12–24 months), Parmigiano Reggiano DOP Extra (ripened ca. 18 months, and used for National and European markets), and Parmigiano Reggiano DOP Export (ripened ca. 18 months, and used for non-European markets). Besides the indication of the ripening period, consumers may have information on the cheese age also based on the stamps proposed by the Consortium: lobster (over 18 months), silver badge (over 22 months), and gold badge (over 30 months).

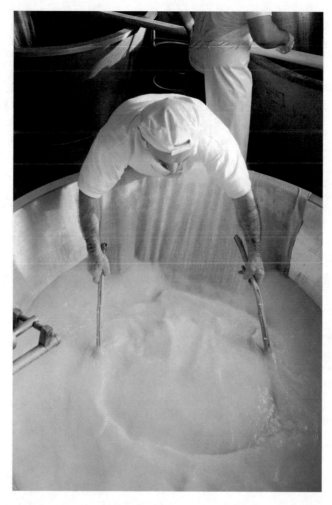

Fig. 6.7 Parmigiano Reggiano cheesemaking. Curd extraction. The photo has been kindly supplied by the Consorzio Parmigiano Reggiano PDO

6.2.1.2 Grana Padano

PDO Grana Padano cheese is manufactured in several provinces of the Po Valley. These include the territory of the Piedmont regions, Lombardy (excluding the part of the province of Mantova on the right of the river Po), Veneto (excluding the province of Belluno), Trento in the Trentino Alto Adige region, Bologna, on the right side of the river Reno, Ferrara, Forlì-Cesena, Piacenza, Ravenna, and Rimini in the Emilia–Romagna region, and some municipalities (Anterivo, Lauregno, Proves, Senale-SanFelice, and Trodena) in the province of Bolzano (www.granapadano. com; Mipaaf 2017; Gobbetti and Di Cagno 2017; AFIDOP-IGP 2016; Gatti et al. 2014; Nevini and Gatti 2013; Mucchetti and Neviani 2006; Mucchetti et al. 1998;

Fig. 6.8 Parmigiano Reggiano cheesemaking. Curd extraction. The photo has been kindly supplied by the Consorzio Parmigiano Reggiano PDO

Neviani and Carini 1994; Parisi 1966) (Figs. 6.20, 6.21, 6.22, 6.23, 6.24, 6.25, 6.26, 6.27, 6.28, 6.29, 6.30, 6.31, 6.32, 6.33, and 6.34).

Several major features distinguish Grana Padano from Parmigiano Reggiano. For Grana Padano, the feeding of high-quality silage fodder is allowed, and the cheese is produced from two consecutive milkings, which are stored at a temperature not less than 8 °C on the farm. The use of silage favors the contamination of raw milk by spore-forming clostridia. To inhibit the eventual late blowing of cheese, caused by butyric fermentation, the addition of lysozyme to the vat milk (20–25 ppm) is allowed. The milk is skimmed by creaming in *bacinelle* or very large tanks (300–500 hL) for ca. 12 h at 8–20 °C (Fig. 6.22). The total microbial count of the milk after holding in the *bacinelle* is low (ca. 10^3–10^4 cfu/mL) compared to that

Fig. 6.9 Parmigiano Reggiano cheesemaking. Curd extraction. The photo has been kindly supplied by the Consorzio Parmigiano Reggiano PDO

of milk for Parmigiano Reggiano, and this is due to the lower temperature of creaming (Mucchetti and Neviani 2006; Neviani and Carini 1994). The fat content of the vat milk for Grana Padano is ca. 2.1–2.2%. As described previously for Parmiggiano Reggiano cheese, creaming in *bacinelle* causes a slight microbial acidification, which may favor rennet activity and the growth of natural whey culture. Currently, several dairies also use raw milk from a single daily collection. This change has stimulated also an innovation in the milk fat standardization. Larger tanks, having a capacity of ca. 10,000 kg, replaced the traditional *bacinella*. These new tanks are easier to clean and the temperature of milk is regulated according to its microbiological quality, rather being dependent on seasonal climatic variations. Besides, the variability of the fat content of each vat decreases. The natural whey culture used as starter is similar to that described for Parmigiano Reggiano. It is prepared from whey from the previous cheesemaking, which is held at a temperature gradient (from ca. 50 °C to ca. 35 °C) for ca. 24 h. Also in this case, the microbial composition of the natural starter is very complex and dominated by thermophilic lactic acid bacteria such as *Lactobacillus helveticus*, *Lactobacillus delbrueckii* subsp. *lactis*, *Lactobacillus delbrueckii* subsp. *bulgaricus*, and *Lactobacillus fermentum* (Gatti et al. 2014; Neviani and Gatti 2013; Rossetti et al. 2008; Santarelli et al. 2008; Gatti et al. 2003; Giraffa and Neviani 1999; Neviani and Carini 1994). Similar to that observed for Parmigiano Reggiano the ratio of obligately homofermentative to heterofermentative species is ca. 10:1 or higher. A large amount of the natural whey culture, ca. 3% (v/v), is added to the vat milk, giving a total titratable acidity of ca. 28–32°SH/50 mL. The use of commercial/selected starters is not allowed. The use of starters, previously selected from the autochthonous bacteria populating the natural whey culture, is permitted, for a maximum of twelve times per year, only when a dramatic decrease of the acidifying power by the natural whey culture is observed.

Fig. 6.10 Parmigiano Reggiano cheesemaking. Curd extraction. The photo has been kindly supplied by the Consorzio Parmigiano Reggiano PDO

The calf rennet (powdered preparation) used for Grana Padano contains less than 3–4% pepsin, and coagulation occurs within 8–12 min. After coagulation, the coagulum is broken into particles having the size of rice grains and cooked at 53–56 °C under stirring. The time from rennet addition at 32–34 °C to the end of cooking is 22–23 min. The combination of heating and acidifying activity by the natural whey cultures allows the formation of the right texture of curd granules and whey drainage (Figs. 6.23, 6.24, 6.25, and 6.26). After cooking, the curd granules settle to the bottom, where they congeal in ca. 30–50 min at a whey temperature not exceeding 52–55 °C (Fig. 6.27). Then, the curd is removed and cut into two portions (Figs. 6.28 and 6.29). Each of these two portions is placed into a mold to give the final shape

Fig. 6.11 Parmigiano Reggiano cheesemaking. Curd extraction. The photo has been kindly supplied by the Consorzio Parmigiano Reggiano PDO

(Figs. 6.30 and 6.31). After 2 days of molding, the cheese is salted by immersion in saturated salt brine (18–22°Bé) for about 3 weeks (Fig. 6.32). The vats used for the manufacture are similar to those used for Parmigiano Reggiano cheese and have a capacity of 10–12 hL and, traditionally, have the shape of an inverted bell made of copper or with an internal copper coating (Figs. 6.23 and 6.24). From each vat, two cheeses, each weighing 35–37 kg after ripening, are produced. Ripening varies from a minimum period of 9 months to more than 20 months (Figs. 6.33 and 6.34).

Grana Padano has a cylindrical shape with a diameter of 35–45 cm and height of 18–25 cm. One form of Grana Padano DOP has a weight in the range 24–40 kg. The texture is hard and the cheese has a very low moisture content and typical compact texture, with or without many very small eyes. Cheese paste shows a minute grain structure and is white or straw-colored. The greater the fat content increases the more the texture tends to become elastic. Aroma and flavor of the paste are fragrant and delicate (not spicy). Also for Grana Padano aroma and flavor are the result of very slow ripening, during which proteolysis is the main biochemical event (AFIDOP 2016; Gatti et al. 2014; Neviani and Gatti 2013; Manzi et al. 2007; Sforza et al. 2004). The rind is hard and smooth and has a natural straw color. The thickness of the cheese rind is ca. 4–8 mm.

The cheeses on the market are the following types: Grana Padano DOP (ripened for at least 9 months), Grana Padano DOP oltre 16 mesi (ripened over 16 months), Grana Padano DOP Riserva (ripened for at least 20 months), and Grana Padano DOP Trentin Grana (manufactured in the province of Trento using a cheesemaking procedure similar to that used for Parmigiano Reggiano cheese) (Mipaaf 2017; Mucchetti and Neviani 2006).

Fig. 6.12 Parmigiano Reggiano cheesemaking. Curd molding. The photo has been kindly supplied by the Consorzio Parmigiano Reggiano PDO

6.2.2 Bagòss

Bagòss is an ancient artisanal cooked cheese manufactured in mountain areas (AFIDOP 2016; Mucchetti and Neviani 2006). Typically, manufacture of the cheese is in the territory of the village of Bagolino (close to Brescia, Lombardy region). The name of the cheese derives from the dialectal name of the inhabitants of Bagolino (Figs. 6.35 and 6.36). Probably, the origin of Bagòss dates back to the sixteenth century, when Bagolino was a country at the border of the Republic of Venice. At that time, Venice dominated the sea. One of the main distinguishing features of Bagoss cheese is the addition of saffron. This is not a local product but a spice very common in Venice, when Venetians traded throughout the world.

Fig. 6.13 Parmigiano Reggiano cheesemaking. Curd molding. The photo has been kindly supplied by the Consorzio Parmigiano Reggiano PDO

Fig. 6.14 Parmigiano Reggiano cheesemaking. Curd molding. The photo has been kindly supplied by the Consorzio Parmigiano Reggiano PDO

No refrigeration of the milk occurs after milking. Raw milk, from one or two consecutive milkings, undergoes partial skimming, according to a creaming protocol similar to that described for Parmigiano Reggiano and Grana Padano cheeses. During winter, the temperature of milk storage varies from 1 to 10 °C, and during the other seasons of the year it depends on the climate. The evening milk may be used without creaming. Starters are not used. A low amount of calf rennet powder (1–2 g/100 kg of milk) is used. The coagulation of milk occurs at 35–40 °C in 45–85 min. Curd breaking to the final dimensions of rice grains occurs progressively by means of different artisanal tools. Breaking is not a continuous process but

Fig. 6.15 Parmigiano Reggiano cheesemaking. Curd salting in brine. The photo has been kindly supplied by the Consorzio Parmigiano Reggiano PDO

Fig. 6.16 Parmigiano Reggiano cheesemaking. Curd salting in brine. The photo has been kindly supplied by the Consorzio Parmigiano Reggiano PDO

has several interruptions, during which curd grains accumulate to the bottom of the vat. The complete rupture operation may last 30–80 min, with the temperature decreasing slightly to 32–35 °C. After cutting, the curd grains are heated to 47–51 °C for 20–34 min. After cooking, curd grains are subjected to stirring under whey for another 4–5 min, without a further increase of temperature. Then, curd grains are held under whey for ca. 30 min. This typical protocol of curd cutting and cooking leads to a significant release of fat into the whey. Molded curd is pressed for 5–6 h and left to drain for 24–48 h. After 24 h, the pH ranges between 5.5 and 5.7, then it remains stable throughout salting and tends to increase slightly during ripening. The absence of starters avoids excessive acidification of the cheese. The curd is

Fig. 6.17 Parmigiano Reggiano cheesemaking. Curd ripening. The photo has been kindly supplied by the Consorzio Parmigiano Reggiano PDO

Fig. 6.18 Parmigiano Reggiano cheesemaking. Curd ripening. The photo has been kindly supplied by the Consorzio Parmigiano Reggiano PDO

dry-salted with continuous treatments for 35–40 days. Usually, cheese ripening takes one year. The cheese microbiota is very complex, including wild thermophilic streptococci, mesophilic lactobacilli, mesophilic enterococci, yeasts, and molds.

The cheese is cylindrical in shape, 10–12 cm high and 40–55 cm in diameter, and weighs 14–22 kg. The cheese paste has a compact granular texture, sometimes with few very small eyes. The color of the cheese paste ranges from strawy-yellow to brown-yellow. The rind is red-brown. Taste and flavor are pronounced and moderately piquant.

Fig. 6.19 Parmigiano Reggiano cheesemaking. Curd ripening. The photo has been kindly supplied by the Consorzio Parmigiano Reggiano PDO

6.2.3 Bitto

Bitto is an artisanal PDO cheese (Mipaaf 2017; AFIDOP 2016; Morandi et al. 2011; Manzi et al. 2007; Mucchetti and Neviani 2006; Mucchetti et al. 1998; Neviani and Carini 1994; Parisi 1966). It is a pressed, cooked, medium-to-long ripened, semi-hard cheese produced from raw cow's milk alone, derived from traditional breeds from the production area, or from a combination with raw goat's milk that does not exceed 10% of the total milk. Typically, Bitto cheese is manufactured in Northern Italy (Lombardy region), especially in the mountain territories of the Valtellina but also in Val Brembana (close to Brescia and Lecco) (Figs. 6.37 and 6.38).

Fig. 6.20 Grana Padano cheese. The photo has been kindly supplied by AFIDOP, Italy

After milking, the milk is not refrigerated and, without creaming, used for cheese manufacture. Cheesemaking is in a copper vat, using direct firewood. Starters are not used. Consequently, the wild cheese microbiota may include thermophilic streptococci, mesophilic lactobacilli, enterococci, and yeasts. Calf rennet is added at 35–39 °C, and coagulation takes place in ca. 30 min. Curd breaking lasts ca. 30 min and particles the size of rice grains are obtained. During cutting, the curd grains are heated to 48–52 °C, and then held and stirred for 15–30 min in whey without further heating. The molded curd is pressed for 10–24 h and left to drain for a further 24–48 h. After 24 h, the pH of the cheese is close to 6.0. Excessive acidification of the curd is avoided because of starters are not used. The cheese is dry-salted using repeated treatments for 8–25 days, which depend on the cheese dimension. Bitto becomes a hard cheese variety after a very prolonged time of ripening. The cheese may be matured for 70 days, but aging for several years also occurs, without altering the sensory and texture characteristics of the cheese. After at least one year of ripening, the cheese may be used grated. Traditionally, some cheeses were ripened for more than 10 years. At the end of ripening, the fat content must be not less than 45% of dry cheese matter and moisture close to 38%.

The cheese is cylindrical in shape, 8–12 cm high and 30–50 cm in diameter, and the weight may vary from 8 to 25 kg. The color of cheese paste ranges from strawy-yellow to brown-yellow, depending on the ripening time. The cheese paste has a compact granular texture, sometimes with few very small holes. The rind is yellow-brown rind with a thickness of ca. 2–4 mm. The flavor is sweet that becomes more pronounced in long-ripened cheeses or depending on the amount of goat's milk used.

Fig. 6.21 Geographical area for the manufacture of Grana Padano cheese

6.2.4 *Nostrano Valtrompia*

Nostrano Valtrompia is a DOP semi-fat cheese, having an extra-hard paste (Mipaaf 2017; Mucchetti and Neviani 2006; Mucchetti et al. 1985, 1993). Partially skimmed raw cow's milk is used with the addition of saffron. The area for cheese manufacture includes some territories of the Province of Brescia (Lombardy Region), particu- larly the Valletrompia Valley and the Gussago Mountains (Figs. 6.39 and 6.40).

Raw milk from a maximum of four consecutive milkings is used. Partial skim- ming takes place in little round basins with a capacity of 25–125 L, according to a creaming protocol similar to that described for Parmigiano Reggiano and Grana Padano cheeses. The full-fat milk from the last milking could be added to skimmed milk from the previous milking. The temperature for skimming ranges from 4 to

Fig. 6.22 Grana Padano cheesemaking. Cheese milk skimming by natural creaming. The photo has been kindly supplied by the Consorzio Grana Padano PDO

Fig. 6.23 Grana Padano cheesemaking. Cheese milk placed in special vat (inverted bell form). The photo has been kindly supplied by the Consorzio Grana Padano PDO

22 °C. This wide interval depends on seasonal variability and the alternation of temperature between day and night; the creaming protocol lasts up to 48 h. Copper vats are used. The traditional protocol for manufacture excludes the use of natural or commercial/selected starters but currently 2% (v/v) of a natural whey culture is allowed. Traditionally, heating of the milk to the coagulation temperature, 36 or 40°C, is by using direct firewood. Liquid calf rennet is used for coagulation, which is complete in 10–30 min. If not added previously, saffron is added during curd breaking (0.05–0.2 g/100 kg of milk). Curd breaking occurs in different steps to reach the dimensions of rice grains. Cooking of curd grains is at 47–52 °C for

Fig. 6.24 Grana Padano cheesemaking. Cheese milk warming. The photo has been kindly supplied by the Consorzio Grana Padano PDO

Fig. 6.25 Grana Padano cheesemaking. Curd breaking. The photo has been kindly supplied by the Consorzio Grana Padano PDO

8–30 min. After cooking, the curd grains are held under the whey for 15–60 min. Molded curd is left to drain for a maximum of 24 h. A dry salting procedure is used, treatments are repeated for 5–20 days, depending on the size of the cheese. The raw milk mesophilic lactobacilli and enterococci are the predominant microbiota, but thermophilic lactic bacteria may contribute to cheese ripening if a natural whey culture is used as starter. The cheese is ripened for at least 12 months and even up to 2 years. The final fat content in the cheese must be within 18–28% in the cheese dry matter. The moisture content must not exceed 36%. During ripening, the cheese rind is treated with linseed oil to prevent fungal contamination.

Fig. 6.26 Grana Padano cheesemaking. Final part of curd breaking. The photo has been kindly supplied by the Consorzio Grana Padano PDO

Fig. 6.27 Grana Padano cheesemaking. Broken curd under warm whey. The photo has been kindly supplied by the Consorzio Grana Padano PDO

The cheese shape is cylindrical with a diameter of 30–45 cm, height of 8–12 cm, and weight of 8–18 kg. The cheese paste is hard, but not too granular and may have uniformly distributed very small holes. The color is straw yellow with a tendency to green-yellow. The rind is tough and has a color varying from brown to reddish-yellow. The cheese has an intense taste, bitterness is absent or slightly perceived, while notes of spicy appear in long-ripened cheeses.

Fig. 6.28 Grana Padano cheesemaking. Curd extraction from the vat. The photo has been kindly supplied by the Consorzio Grana Padano PDO

Fig. 6.29 Grana Padano cheesemaking. Curd drainage. The photo has been kindly supplied by the Consorzio Grana Padano PDO

6.2.5   Silter

Silter is a DOP hard cheese produced exclusively from partially skimmed raw cow's milk (Mipaaf 2017; Mucchetti and Neviani 2006; Bianchi Salvatori and Sacco 1981; Delforno and Losi 1975; Corradini et al. 1973). The name Silter comes from the traditional name of the ripening rooms. Probably, it refers to a term of Celtic origin, which corresponds to the Italian term *Casera*. The area for cheese

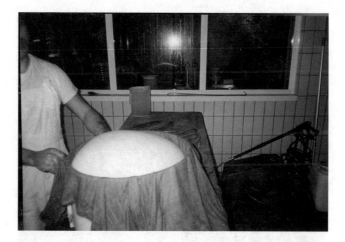

Fig. 6.30 Grana Padano cheesemaking. Curd molding. The photo has been kindly supplied by the Consorzio Grana Padano PDO

manufacture includes the entire territory of the province of Brescia (Lombardy region) and, in particular, the Camonica Valley and Sebino territory (Figs. 6.41 and 6.42).

No refrigeration of the milk occurs after milking. Raw milk, from one or two consecutive milkings, undergoes partial skimming, according to a creaming protocol similar to that described for Parmigiano Reggiano and Grana Padano cheeses. The use of natural milk or whey cultures or commercial/selected starters is optimal. Milk coagulation is by using liquid or powdered calf rennet at 36–40 °C, and is complete in 25–45 min. The curd is broken progressively, with a sequence of ruptures giving a final dimension of rice or corn grains. After cutting is completed, heating of the curd grains occurs at 46–52 °C. When the cooking procedure is finished, the curd grains remain under whey for further 20–60 min without further warming. The molded curd is pressed gently, allowing drainage for 12–24 h. A dry salting procedure is used, treatments are repeated for 4–12 days. The cheese microbiota is strictly related to raw milk microbiota (mainly mesophilic lactobacilli, enterococci) and to thermophilic lactic acid bacteria if a natural or commercial/selected starter is used. Typically, the cheese is ripened for at least 100 days, but maturation could continue even for 2 years. The fat content of the ripened cheese must be within the large interval of 27–45% of the dry cheese matter. The moisture content does not exceed 40%.

The cheese shape is cylindrical with a height between 8 and 10 cm, and a diameter between 34 and 40 cm, and the weight ranges from 10 to 16 kg. Usually, the cheese with higher weight is chosen for the longest ripening time. The cheese paste is tough, never too elastic, sometimes with uniformly distributed small holes. The color varies from intense yellow to white, depending on the seasonal feeding of cows and on the ripening time. The rind has a pale yellow color, tending to brown after ripening and as a consequence of oiling. The cheese has a sweet taste, bitterness is

Fig. 6.31 Grana Padano cheesemaking. Curd molding. The photo has been kindly supplied by the Consorzio Grana Padano PDO

absent or slightly perceived, while notes of spicy appear in long-ripened cheeses. The smell and the flavor are persistent; among the most often perceived are dried fruit, butter and chestnut.

6.2.6   Asiago Cheeses

Asiago are low-cooked PDO cheese varieties made from cow's milk (Mipaaf 2017; AFIDOP 2016; Cozzi et al. 2009; Manzi et al. 2007; Mucchetti and Neviani 2006; Merlo 2002; Dalla Torre and Fontanella 1955; Scapaccino 1936). Low cooking

Fig. 6.32 Grana Padano cheesemaking. Curd salting in brine. The photo has been kindly supplied by the Consorzio Grana Padano PDO

means the use of a cooking temperature lower than that used for making Parmigiano Reggiano and Grana Padano cheeses (temperature higher than the renneting temperature but below to 50 °C). Only in the past, these cheeses could be produced using ewe's milk also. The cheese is manufactured in some territories of Vicenza, Padova, Treviso (Veneto region), and Trento (Trentino-Alto Adige region) provinces (Figs. 6.43 and 6.44). Asiago cheeses made in dairies located in mountainous areas, at an altitude of more than 600 m, are identified with the brand *Prodotto della Montagna* (literally, mountain production). Two main varieties are distinguished: Asiago Pressato and Asiago d'Allevo, and each is manufactured as several different types, which differ mainly in the extent of ripening.

6.2.6.1   Asiago d'Allevo

Asiago d'Allevo is a hard PDO cheese produced from partially skimmed raw or thermized cow's milk. The most traditional production of the cheese occurs during summer and comes from the Alps.

Before cheesemaking, the milk is stored at a temperature ranging between 4 and 11 °C. Raw milk from one or two consecutive milkings is partially skimmed, according to a creaming protocol similar to that described for Parmigiano Reggiano and Grana Padano cheeses. Natural milk or whey cultures or commercial/selected starters may be used. The coagulation of milk takes place at 33–37 °C for 15–30 min, using liquid or powdered rennet. The protocol for manufacture allows the use of lysozyme, at a maximum level of 2.5 g/100 kg milk. The use of lysozyme is prohibited for making the type Asiago d'Allevo *Prodotto della Montagna*. In some cases, the protocol also allows the addition of 0.5% NaCl to milk. The curd is progressively broken into grains with dimension of ca. 0.5 cm, close to the size of a

Fig. 6.33 Grana Padano cheesemaking. Early curd ripening. The photo has been kindly supplied by the Consorzio Grana Padano PDO

hazelnut. The curd is broken at 46–48 °C in ca. 10–15 min. After breaking, the curds may be held under warm whey for 20–30 min. After removal from the vat, the molded curd could be pressed softly to promote whey drainage and maintained in a warm room for few hours. Salting may be done by using dry salt for a period of 10–20 days at 12–16 °C or in brine (16–21°Bé) for ca. 10 days.

The cheese shape is cylindrical, 9–12 cm high and 30–35 cm in diameter, and weighs 8–12 kg. The cheese paste has a compact structure with small and medium-sized holes, and becomes granular in cheese ripened for over a year. The color of the cheese pasta is slightly straw-yellow in young cheese but becomes deeper until amber after long ripening. The rind is smooth and regular. The flavor is sweet for young cheeses but it can become pleasantly spicy after long ripening.

Fig. 6.34 Grana Padano cheesemaking. Curd ripening. The photo has been kindly supplied by the Consorzio Grana Padano PDO

Fig. 6.35 Bagòss cheese. The photo has been kindly supplied by Prof. Germano Mucchetti (Parma University, Italy)

Several types of Asiago d'Allevo are present on the market: the Mezzano cheese ripened for at least 4 months; the Vecchio and Stravecchio cheeses ripened for 10 and 15 months, respectively; and the *Prodotto della Montagna* cheese ripened for at least 90 days.

Fig. 6.36 Geographical area for the manufacture of Bagòss cheese

6.2.6.2   Asiago Pressato

Asiago Pressato cheese has a DOP recognition. The manufacture is from full-fat raw or pasteurized cow's milk.

Before cheesemaking, the milk is stored at a temperature ranging from 4 to 11 °C. Currently, the use of pasteurized milk is preferred. A thermophilic commercial/selected starter, consisting mainly of *Streptococcus thermophilus and Lactobacillus delbrueckii subsp. bulgaricus* strains, is added to the vat milk. Traditionally, natural milk cultures have been used but rarely nowadays. The milk is coagulated at 35–40 °C in 15–25 min, using liquid or powdered rennet. In some cases, the protocol for cheesemaking allows the use of NaCl at 0.25–0.29%. The curd is broken into grains with dimensions of ca. 1 cm, about the size of a walnut. The broken curd is cooked at 42–46 °C in 10–15 min, using an increasing gradient

Fig. 6.37 Bitto cheese. The photo has been kindly supplied by AFIDOP, Italy

of temperature. During cooking, warm water may be added to remove some lactose, and to reduce acidification of the curd during the early stage of ripening. After removal from the vat, the broken curd may be washed with water to further reduce acidification and pressed for 3–12 h to allow whey drainage. The pressed cheese is then placed for 2–6 days in rooms at 12–20 °C for the so-called *frescura*. During this period, much fermentation occurs. The practice of washing the curd with water and the speed of whey drainage prevents the pH of the cheese falling below 5.2. Salting is done in brine (17–20°Bé) for ca. 3 days or using dry salt. Asiago Pressato cheese is ripened for a minimum of 20 days, rarely exceeding 40 days. The temperature of the ripening room must be between 10 and 15 °C. The type *Prodotto della Montagna* is ripened for at least 30 days, and the values of temperature and humidity for maturation are those determined by the natural environmental conditions.

The cheese shape is cylindrical, 11–15 cm high and 30–40 cm in diameter, and weighs 11–15 kg. The cheese paste shows marked and irregular holes, because of curd processing and fermentation. The color of the cheese paste varies from white to slightly straw-colored. The rind is thin and elastic. The flavor and the taste are delicate and pleasant.

6.2.7 Branzi

Branzi is a low cooked and pressed cheese with semi-soft or semi-hard texture, depending on the duration of ripening (Mucchetti and Neviani 2006; Delforno and Fondrini 1966). Low cooking means the use of a cooking temperature lower than that used for making Parmigiano Reggiano and Grana Padano cheeses. Currently, Branzi is manufactured in the mountains around Bergamo city (Alta Valle Brembana

Fig. 6.38 Geographical area for the manufacture of Bitto cheese

and Valle Cavallina) (Figs. 6.45 and 6.46) from whole cow's milk (sometimes partially skimmed and/or supplemented with sheep's milk and/or goat's milk). In the alpine dairies, the manufacture is limited to the summer period.

Traditionally, only raw whole milk is used. Natural milk or whey cultures or commercial/selected thermophilic starters may be used but not for making the traditional cheese. Coagulation by calf rennet takes 30–45 min at 34–36 °C. After cutting the coagulum to 3–5 mm grains, the curd grains undergo cooking at 43–46 °C for 25–30 min. After resting for some minutes under whey, the molded curd is pressed. Salting is carried out in brine (17–18°Bé) for 3 days or using dry salt for a total of 10–15 days. Ripening varies over a wide range of time, from 1 to 7 months at ca. 8–16 °C and an environmental humidity of ca. 85–90%.

The cheese has a cylindrical shape with a diameter of 40–50 cm, a height of 8–10 cm and weighs 8–15 kg. The cheese paste has a soft/semi-soft structure that

Fig. 6.39 Nostrano Valtrompia cheese. The photo has been kindly supplied by Prof. Germano Mucchetti (Parma University, Italy)

becomes more elastic during ripening. The paste may contain little holes. The color of the cheese paste ranges from pale yellow to golden-yellow. The cheese rind is thin and yellow-brown. The flavor is sweet for young cheeses and may become little spicy for ripened cheeses.

6.2.8   Fontina

Fontina is a DOP low-cooked cheese, obtained from raw whole cow's milk of Valdostana breeds, pressed and medium–long ripened (Mipaaf 2017; AFIDOP 2016; Bontempo et al. 2011; Berard et al. 2007; Manzi et al. 2007; Mucchetti and Neviani 2006; Pinarelli 2006; Mantovani et al. 2003; Pellegrino et al. 1995; Battistotti et al. 1976; Carbone 1957; Strozzi 1941; Besana 1886). Low cooking means the use of a cooking temperature lower than that used for making Parmigiano Reggiano and Grana Padano cheeses. The manufacture is only in the Valle d'Aosta region (Figs. 6.47 and 6.48). The breed of the livestock and the feeding with local fodder are the main links with the territory.

   The milk used is from a single milking without refrigeration, which is prohibited; milk thermization or pasteurization are also prohibited. Cheesemaking uses only raw cow's milk. Before renneting, the milk should not have undergone warming above 36 °C. Traditionally, no starter is used and milk acidification occurs only based on the metabolic activity of the autochthonous milk microbiota. The milk is renneted using powdered or liquid calf rennet at 34–37 °C for 30–65 min. The curd is broken progressively and using a variable temperature. Cooking reaches the temperature of 48–52 °C over a total of ca. 60 min. After holding for 10 min under the whey, the curd grains are molded and pressed for ca. 12 h. The pH of the curd and fresh cheese usually does not fall below 5.2 and rises to ca. 6.0 during ripening. The cheeses are salted using dry salt, either manually or mechanically for a period of

Fig. 6.40 Geographical area for the manufacture of Nostrano Valtrompia cheese

1–2 months. The traditional ripening period is 3 months at 6–12 ° C and a relative humidity higher than 90%. Frequently, ripening occurs in natural, humid, and cold caves. Under these conditions, the temperature is almost constant at 7–8 °C and the relative humidity reaches a very high level. The ripening in the caves allows the cheese to be ripened even for 10 months yet maintaining the softness of the paste. During the first month of ripening, the cheeses are salted in the morning and wiped with whey or *scotta* (a by-product from Ricotta making) in the evening and turned. Prolonged ripening for up to 7 months leads to an increase in the quantity of free amino acids.

The cheese has a cylindrical shape and a characteristic concave form, with a diameter of 35–45 cm, a height of 7–10 cm, and a weight of 7.5–12 kg. The cheese paste has an elastic and soft structure depending on the production period and is soft

Fig. 6.41 Silter cheese. The photo has been kindly supplied by ONAF (Italian Cheese Taster Organisation, www.onaf.it)

in the mouth. It contains characteristic few and heterogeneously distributed holes. The color of the cheese paste varies from ivory to straw yellow. The rind is thin, soft, or semi-soft depending on the length of ripening, and has a compact structure, light brown to dark in color according to the ripening time and ripening environmental conditions. The flavor is more or less intensely sweet and delicate.

Fig. 6.42 Geographical area for the manufacture of Silter cheese

6.2.9   Formai de Mut dell'Alta Val Brembana

Formai de Mut is a PDO low-cooked cheese made from cow's milk (Mipaaf 2017; AFIDOP 2016; Paleari et al. 1991). Manufacture takes place on the alpine high mountains of the Brembana Valley, and in the Orobic Alps in the province of Bergamo (Lombardy region) (Figs. 6.49 and 6.50).

The cow's milk comes from one or two daily milking of cattle fed mainly with green fodder produced in the area established by internal regulation. Whole raw milk is used without addition of natural milk or whey cultures or commercial/ selected starters. After the addition of rennet, milk coagulation occurs in ca. 30 min at a temperature between 35 and 37 °C. The first break of the curd is followed by

Fig. 6.43 Asiago cheese. The photo has been kindly supplied by AFIDOP, Italy

curd cooking for ca. 30 min at 45–47 °C. The curd is held under the warm whey for ca. 20 min. Pressing of molded curd is used and the curd is left to drain whey for a maximum of 48 h. Salting is done either using dry NaCl or brine and is repeated for a number of times (8–12) depending on cheese size. The minimum ripening lasts 40 days but it may be extend up to 1 year. The minimum fat content is 45% on the cheese dry matter.

The cheese shape is cylindrical with a diameter ranging from 30 to 40 cm, a height of 8 to 10 cm. and a weight of 8 to 12 kg. The cheese paste has a compact ivory color with more or less homogeneously distributed holes. The rind is thin and compact with a light yellow color tending to gray with seasoning. The flavor is delicate, not spicy and nonfermented.

6.2.10 Montasio

Montasio PDO is a hard, low-cooked cheese produced from cow's milk (Mipaaf 2017; AFIDOP 2016; Romanzin et al. 2013; 2015; Manzi et al. 2007; Mucchetti and Neviani 2006; Innocente 1997; Di Bidino 1979; Dalforno 1970; Braidot 1948; Besana 1915; Tosi 1904). The cheese derives its name from the eponymous place in the Julian Alps. Currently, its manufacture extends to the Friuli region and to several provinces of the Veneto region (Belluno, Treviso, Padova, and Venezia) (Figs. 6.51 and 6.52). The cheese is produced throughout the year, but traditionally it is most appreciated during summer in the mountains.

A mixture of cow's milk from two consecutive milkings is used; skimming of the milk from the evening milking may occur. The milk may be stored before cheesemaking at a temperature not lower than 4 °C. Traditionally, only raw milk is

Fig. 6.44 Geographical area for the manufacture of Asiago cheese

used. The milk must not be pasteurized and must show a clear positivity in the phosphatase assay. This prescription allows the possibility of thermization of milk. The use of lysozyme is permitted. Contrarily to the past, currently natural whey cultures or commercial/selected thermophilic starters may be used. Starters are added to the milk at 32–36 °C and coagulation by calf rennet takes place in 30–40 min. The curd is broken into particles having the size of rice grains. Cooking is at 42–48 °C for more than 10 min. Molded curd is pressed for 24 h and salted using dry NaCl or by immersion in saturated brine (18–20°Bè), or a mixture of both these treatments. Ripening is at a temperature not less than 8 °C for the first 30 days and thereafter at a higher temperature, usually close to 18 °C, at a humidity of ca. 80%. Ripening lasts for at least 2 months. At the end of ripening, the cheese must have a minimum fat content of 40% in the cheese dry matter and a maximum moisture content of ca. 36%.

Fig. 6.45 Branzi cheese. The photo has been kindly supplied by Latteria Branzi, Italy

The cheese has a cylindrical shape with a diameter of 30–35 cm, a height of 8 cm or less, and a weight of 6–8 kg. The cheese paste has a granular texture with very small holes, which are considered a typical feature. The color is white or slightly straw-yellow in the young cheese and becomes deeper straw-yellow and eventually amber after long ripening. The cheese rind is brown, smooth, and elastic. The flavor is sweet for young cheeses and may become pleasantly spicy and moderately piquant after long ripening.

Based on the duration of ripening, the cheese has the following types on the market: Montasio Fresco (ripening of 2–4 months); Montasio Mezzano or Semistagionato (4–8 months); Montasio Stagionato (more than 10 months); and Montasio Stravecchio (more than 18 months). The brand *Prodotto della Montagna* could be used if the cheese is produced in particular mountain areas.

Fig. 6.46 Geographical area for the manufacture of Branzi cheese

6.2.11   Monte Veronese Cheeses

Monte Veronese PDO cheese is produced in the Northern part of the province of Verona in the Veneto region (Mipaaf 2017; AFIDOP-IGP 2016; Manzi et al. 2007; Mucchetti and Neviani 2006; Maggi and Ballerini 1994). It is a low-cooked cheese produced according to either of two procedures: whole cow's milk Monte Veonese cheese (table cheese) and partially skimmed milk cheese, in the case of Monte Veronese "d'Allevo" (table or grated cheese) (Figs. 6.53 and 6.54).

Fig. 6.47 Fontina cheese. The photo has been kindly supplied by AFIDOP, Italy

6.2.11.1   "Whole Milk" Monte Veronese

"Whole milk" Monte Veronese cheese is produced from whole cow's milk, obtained from one or two consecutive milkings. The whole raw milk is refrigerated and pasteurized. The addition of natural milk or whey cultures or commercial/selected starters is permitted. The milk is coagulated using calf rennet, usually at 33–34 °C in 15–20 min. The curd is cut into particles having a size of rice grains. Curd grains are heated at 43–45 °C for ca. 10 min. Then, the curd stays in the vat under whey for ca. 25–30 min, before being divided and molded. About 24 h after molding, the curd is salted using dry salt or brine. The cheeses are ripened for 30 days, not less than 25 days.

The cheese has a cylindrical shape, with nearly flat faces and slightly convex sides, with a diameter of 25–35 cm and a height of 6–10 cm. The weight varies from 6 to 9 kg. The cheese rind and paste are yellow more or less intensely, depending on the production period. The holes are scattered and usually have the size of 2–3 mm. The cheese has a delicate and pleasant taste of fresh milk, cream, or fresh butter. The flavor is sweet, slightly acidic like yogurt. The cheese has a soft, easily soluble structure in the mouth.

6.2.11.2   Monte Veronese "d'Allevo"

Monte Veronese "d'Allevo" is produced from partially skimmed cow's milk, coming from one or two consecutive milkings. Raw milk is refrigerated and pasteurized. The addition of natural milk starter or specific starters (if produced in different dairies of the PDO zone) is permitted. The milk is coagulated using calf rennet at a

Fig. 6.48 Geographical area for the manufacture of Fontina cheese

temperature of 33–34 °C for 25–30 min. The curd is cut into pieces the size of a rice grain. After cutting, the curd grains are heated at 46–48 °C for ca. 15 min. Then, curd grains are held for a further 25–30 min in the vat under whey. About 24 h after portioning and molding, the curd is salted using dry salt or brine. Ripening lasts for a minimum of 90 days, if the cheese is used as a table cheese, or for a minimum of 6 months if it is used as a grated cheese. Ripening could extend to 1 or 2 years.

The cheese has a cylindrical shape, with nearly flat faces and slightly convex sides, with a diameter ranging from 25 and 35 cm and a height between 7 and 11 cm. The cheese weighs between 7 and 10 kg. The cheese paste is more consistent than that of "whole milk" cheese, slightly elastic when 3 months old; it becomes tough and slightly grainy with ripening. The smell resembles the pleasant smell of ripe or cooked butter, but also hay and herbs like sage. Compared to the "whole milk" type, the cheese has a tastier flavor typical of ripened cheese. The flavor resembles that of ripe butter and hazelnuts. It tends to become slightly spicy during ripening.

Fig. 6.49 Formai de Mut cheese. The photo has been kindly supplied by ONAF (Italian Cheese Taster Organisation, www.onaf.it)

6.2.12   *Ossolano*

Ossolano PDO is a low-cooked hard cheese (Mipaaf 2017; Mucchetti and Neviani 2006; Zeppa et al. 2002; Gorini 1936). The territory for cheesemaking corresponds the *ossolano* area, which is close to Monte Rosa and includes some territories close to Verbania Cusio Ossola (Piedmont region) (Figs. 6.55 and 6.56). The DOP cheese variety named *Ossolano d'Alpe* is manufactured only from cow's milk coming from pastures located in the same territory, but at an altitude not less than 1400 m and produced from June 1 to September 30. Raw or pasteurized cow's milk is used, and the use of natural milk or whey cultures or commercial/selected starters is permitted. The use of raw milk is most frequent in the high mountain areas. Usually, milk coagulation takes place at a temperature close to 36–39 °C. Curd breaking requires 5–10 min to get particles of the dimensions of corn grains. Heating of curd grains is at 42–45 °C for 15–30 min. Molded curd is pressed and left to drain for a maximum of 12 h. Salting is performed using either dry salt or brine, and treatments are continued for a long time, depending on cheese size. The cheese is ripened for at least 60 days. The minimum fat content of ripened cheese is 40% in cheese dry matter.

The shape of the cheese is cylindrical with a diameter ranging from 29 to 32 cm and a height from 6 to 9 cm. The weight of the cheese varies from 6 to 7 kg (5–6 kg for Ossolano d'Alpe). The cheese paste is consistent, elastic, with irregular small holes, and the color varies from slightly or intense straw to intense yellow. The rind is smooth straw-colored, tending to become more intense during ripening.

Fig. 6.50 Geographical area for the manufacture of Formai de Mut cheese

The cheese has a characteristic delicate flavor, linked to seasonal flora varieties, becoming more intense and fragrant with aging.

6.2.13   Piave

Piave is a DOP hard cheese obtained from partially skimmed cow's milk (Mipaaf 2017; AFIDOP-IGP 2016). At least 80% of the milk must come from animal breeds which are typical of the production area. The area for cheese manufacture includes the entire territory of the Province of Belluno (Veneto Region) (Figs. 6.57 and 6.58).

Pasteurized milk, inoculated with milk or whey natural cultures and/or commercial/selected starters, is used for cheesemaking. No strict indication of temperature

Fig. 6.51 Montasio cheese. The photo has been kindly supplied by AFIDOP, Italy

and time are given for milk coagulation, depending on the type of the cheese. The most frequent protocol uses ca. 36 °C for 20–40 min. The curd is cut into pieces the size of rice grains. Then, curd grains are heated at 44–47 °C for several minutes. After cooking, the curd grains remain under whey for 1.5–2 h without further warming. Pressing the molded curd lasts 30–40 min, allowing whey drainage for 12 h. The cheese is salted in brine (16–21°Bé) for at least 2 days and ripened at 8–14 °C and 70–90% relative humidity. The cheese microbiota is strictly related to thermophilic lactic acid bacteria from natural or commercial starters. Ripening is variable according to the type of cheese, ranging from a minimum of 20 days to more than 18 months.

The cheese shape is cylindrical with a diameter of 27–32 cm and a height of 7–8 cm. The cheese weight varies depending on ripening time: ca. 7.0 kg for the Fresco type; ca. 6.5 kg for the Mezzano type; ca. 6.0 kg for the Vecchio type; ca. 5.8 kg for the Old Gold Selection; and ca. 5.5 kg for the Old Reserve. The color of the cheese paste is mainly white in the fresh type, becoming of an intense strawberry color in the most ripened cheeses, which are used as grating cheeses also. The cheese paste has a homogeneous structure, characterized by the absence of holes, and in the most ripened types the texture becomes dry, granular, and friable. The cheese taste, initially, is sweet and lactic, especially in the Fresco type. On ripening the taste becomes more sapid and progressively more intense and slightly spicy in the longer-ripened cheeses. The cheese rind is tender and clear in the fresh type, becoming thicker and more consistent on prolonged ripening.

The following types of Piave cheese are on the market: Piave DOP Fresco (20–60 days of ripening), Piave DOP Mezzano (61–180 days), Piave DOP Vecchio (more than 180 days), Piave DOP Vecchio selezione Oro (Old Gold Selection—more than 12 months), and Piave DOP Vecchio Riserva (Old Reserve—more than 18 months).

Fig. 6.52 Geographical area for the manufacture of Montasio cheese

6.2.14   Stelvio Stilfser

Stelvio or Stilfser is a PDO semi-hard cheese produced from cow's milk in some territories of the Bolzano province (Trentino Alto Adige region) (Mipaaf 2017; AFIDOP-IGP 2016). The name indicates the mountainous region of the Stelvio National Park (Figs. 6.59 and 6.60). The particular characteristics of this cheese are due mainly to the typical high mountain vegetation, which is the basis of the animals' feed. The cheese has, over time, maintained the peculiar characteristics of the Alpine environment of origin, made up of the mountain range which represents the center of major production (between 500 and 2000 m of altitude).

The milk used for cheesemaking must come from cows fed with fodder produced within the characteristic territory. The milk must have a minimum fat content of 3.45 and 3.1% protein. It is allowed to cool the milk to ca. 6.0–9.0 °C for a maxi-

Fig. 6.53 Monte Veronese cheese. The photo has been kindly supplied by AFIDOP, Italy

mum of 25 h. Overall, the milk must be used for cheesemaking within 48 h from milking. The milk may be skimmed slightly so that the fat content could be adjusted within the range 3.45 and 3.6%. The milk is High Temperature Short Time (HTST) pasteurized. Lysozyme may be added to milk (up to 2 g per 100 L of milk). Natural milk or whey cultures or commercial/starters starters, cultured in milk produced in Stelvio territory, may be used. The microbiota of these starters is composed mainly of mesophilic lactic acid bacteria that become dominant during cheese ripening. The coagulation of milk takes 20–27 min, using liquid or dry calf rennet at 32–33 °C. The curd is broken in 10–15 min to pieces the size of corn grains. During breaking, the curd is stirred for 20–30 min, then a part of the whey (25–35% of the total volume) is removed and the broken curd is heated by adding a quantity of warm water (50–70 °C) to give a final temperature of 36–40 °C. The time required for the total process, from the addition of rennet to molding, is 80–90 min. Molded curd undergoes a slight pressure for 30–240 min to complete whey drainage. Once the pressing is complete, the curd is left in an air-conditioned room to reach a sufficient degree of acidity. After that, the curd is cooled by immersion in fresh water for 1–3 h. This operation allows adjustment of the fermentative performance by mesophilic lactic acid bacteria and acidification of the curd, avoiding excessive harmful acid that could have a negative impact on the quality of the ripened cheese. Salting uses brine (16–20°Bé) at 12–15 °C for 36–48 h, depending on cheese size. Ripening is at 10–14 °C and a humidity of 85–95%. The percentage of fat in the cheese dry matter is equal to or higher than 50%, and the moisture content does not exceed 44%. The ripening period is for not less than 62 days.

The cheese has a cylindrical shape with flat or nearly flat faces and a straight or slightly concave side. The weight varies from 8 to 10 kg, the diameter ranges between 36 and 38 cm and the height is around 8–10 cm. The cheese paste, with a compact texture and elastic texture, is pale yellow straw-colored, with irregular

Fig. 6.54 Geographical area for the manufacture of Monte Veronese cheese

small and medium-sized holes. The rind must have the typical color, varying from orange yellow to brown orange. The taste and flavor are intensely aromatic.

6.2.15   Toma Piemontese and Toma Piemontse semi-fat

The Toma Piemontese DOP name includes a large number of cheeses produced in the territory of Novara, Vercelli, Biella, Torino, and Cuneo (Piedmont Region) (Mipaaf 2017; AFIDOP-IGP 2016; Revello Chion et al. 2010; Delforno 1975) (Figs. 6.61 and 6.62). Indeed, the name Toma Piemontese recognizes a large variety of traditional cheeses, usually lowcooked, with a short–medium ripening time, obtained from raw or thermized or pasteurized whole or partially skimmed milk.

Fig. 6.55 Ossolano cheese. The photo has been kindly supplied by ONAF (Italian Cheese Taster Organisation, www.onaf.it)

Different starters may be used, but the traditional cheeses were produced without their addition. Therefore, the cheese microbiota is extremely heterogeneous, depending on the type of cheese technology used. Usually, *Streptococcus thermophilus* and mesophilic lactic acid bacteria are present, but in the raw milk cheeses, particularly if ripened for less than 40–50 days, also, coliforms and staphylococci may be present. Renneting is done using liquid calf rennet (rarely rennet paste) at 32–35 °C for 30–40 min. The coagulum is broken into little pieces and subjected to low cooking at ca. 45 °C. After holding for a few minutes under whey, molding of the curd occurs. Whey drainage takes place over about 24 h. The pH of the curd after whey drainage is usually ca. 6.0. The cheeses are salted in brine (16–18°Bé) or using dry salt for a time depending to the cheese size (usually a maximum of 15 days). The ripening time for the large cheeses (more than 6 kg) lasts about 30–60 days, while smaller cheeses may be ripened for 15 days.

The cheese has a cylindrical shape with a wide range of dimensions, 6–12 cm high, 15–35 cm in diameter and a final weight that may vary over a very large range, 1.8 and 8.0 kg. Usually, the cheese paste is elastic with a regular distribution of small eyes. The color of the cheese paste is white-slightly straw yellow. The rind is thin and elastic with a color ranging from yellow to red in long-ripened cheese. The characteristic flavor of different Toma cheeses is sweet.

6.2.16   Valle d'Aosta Fromadzo

Valle d'Aosta Fromadzo or Vallée d'Aoste Fromadzo is a PDO low-cooked cheese manufactured from partially skimmed cow's milk (Mipaaf 2017). A small percentage of goat's milk may be added. The milk for cheesemaking is obtained from the entire territory of the Valle d'Aosta region (Figs. 6.63 and 6.64). Two types of Vallée d'Aoste Fromadzo cheese are produced: the semi-fat cheese (fat content in the dry matter between 20 and 35%) and low-fat cheese (fat content less than 20%).

Fig. 6.56 Geographical area for the manufacture of Ossolano cheese

Usually, raw milk is used and the addition of a commercial/selected starter is permitted. Traditionally, skimming the milk is according to a creaming protocol similar to that described for Parmigiano Reggiano and Grana Padano cheeses. Creaming for the semi-fat type is at room temperature for 12–24 h, while this time for the low-fat type extends to 24–36 h. Usually, the coagulation of the milk is at 34–40 °C. After breaking the curd, the grains are heated at 45 °C for a time varying from 6 to 35 min. A slight pressing of the molded curd occurs. Salting is done by using dry salt or brine, and repeated treatments are used depending on cheese size. During ripening, the cheese rind may be washed with water and brine. This practice is similar to that used for Fontina cheese and it explains the origin of reddish nuances present on the rind of the long-ripened cheese due to the formation of pigments by salt-tolerant bacteria. The ripening period varies from a minimum of 60 days to a maximum of 8–10 months. The short-aged product is used as a table cheese, while

Fig. 6.57 Piave cheese. The photo has been kindly supplied by AFIDOP, Italy

the long-ripened cheese is sometimes consumed after grinding. The cheese may be flavored by the addition of seeds or parts of aromatic plants.

The cheese shape is cylindrical with a diameter ranging from 15 and 30 cm and a height between 5 and 20 cm. The weight varies from 1 to 7 kg depending on the size of the cheese, technological conditions and duration of ripening. The structure of the cheese paste is compact with small and medium-sized holes with a straw color tending to become gray with the advance of ripening. The rind of Fromadzo is straw-colored tending with the prolongation of ripening to gray with possible shades of reddish. The flavor and taste are characteristic, fragrant, and sweet in fresh cheeses and more pronounced, slightly salty, sometimes with a spicy note in long-ripened cheeses. A particular aroma of mountain herbs may be present, especially in cheeses produced in summer.

6.2.17 Valtellina Casera

Valtellina Casera is a DOP low-cooked and pressed cheese made from partially skimmed (semi-fat) cow's milk (Mipaaf 2017; AFIDOP 2016; Mucchetti and Neviani 2006; Merlo 2001; Paleari et al. 1993; Gusmeroli et al. 1988). The milk is from cattle fed with plants and grasses grown in the production area. Manufacture is in the territory of Sondrio province, in the Lombardy region (Figs. 6.65 and 6.66).

Partially skimmed milk is used for manufacture. Skimming at 10–15 °C in a basin or centrifugal skimming could be performed. Before skimming, raw milk may be cooled. Milk could be pasteurized before being coagulated. Natural milk or whey cultures or commercial/selected starters may be added to the vat milk. Renneting is done using calf rennet at 35–37 °C in 35–45 min. The curd is broken for 10–15 min

Fig. 6.58 Geographical area for the manufacture of Piave cheese

to particles of the dimensions of corn grains. The curd grains are cooked at 40–45 °C
for ca. 30 min. At the end of stirring, curd grains are held under the warm whey for
15–20 min before removal. The molded curd is gradually pressed for 8–12 h and
then salted with dry salt or in brine. Salting using dry salt continues with repeated
treatments for 10–16 days, depending on cheese size. Salting in brine (16–20°Bé) is
12–16 °C for 48–72 h, depending on cheese size. The cheese is ripened in special
rooms at 6–13 °C and with a relative humidity of not less than 80%. It is necessary
to differentiate cheeses made from pasteurized milk in which the natural or commercial starter microbiota is predominant, from cheeses made from raw milk with
natural acidification, where the microbiota is typically heterogeneous, as usual in
raw milk cheese. In the ripened cheese, the percentage of fat in the cheese dry matter must be equal to or higher than 34% and the average moisture content at 70 days

Fig. 6.59   Stelvio-Stilfser cheese. The photo has been kindly supplied by AFIDOP, Italy

must be close to 41%. The ripened cheese can also be used grated. The cheese is ripened for a minimum of 70 days.

The cheese has a cylindrical shape with a diameter of 30–45 cm and a height ca. 8–10 cm, and a weight of 7–12 kg. The shape, size and weight may be subject to slight variations depending on the technical conditions for manufacture and the ripening time. The cheese paste, with a compact texture and elastic texture, has a pale yellow and straw color, with irregular small and medium-sized holes. The color varies from white to straw yellow, depending on the production period and ripening. The rind must have the typical color ranging from orange-yellow to brown-orange. The taste and flavor are intensely aromatic. Flavor is sweet, characteristic, with a particular aroma, more intense for long ripening.

6.2.18   Bra

Bra is a PDO cheese produced in the Cuneo territories (Piedmont region) from partially skimmed cow's milk. In some cases, goat's milk and/or ewe's milk, up to a maximum of 20%, may be added to the cow's milk (Mipaaf 2017; AFIDOP 2016; Manzi et al. 2007; Mucchetti and Neviani 2006; Delforno 1958; Fascetti 1903) (Figs. 6.67 and 6.68). There are two different types of Bra cheese, Soft and Hard.

Fig. 6.60 Geographical area for the manufacture of Stelvio-Stifser cheese

6.2.18.1   Soft Bra

The manufacture of soft or semi-soft Bra cheese is from whole cow's milk (some-times supplemented by a small quantity of goat's milk or ewe's milk, not more than 20%) or sometimes from partially skimmed cow's milk (skimming is for 10–18 h at a maximum temperature of ca. 18 °C, or by partial mechanical skimming). Semi-soft Bra cheese is produced without curd cooking.

The hygiene of the raw milk could be improved even using thermal treatments. Traditionally, cheese was produced without starters, but today, natural milk or whey cultures or commercial/selected starters may be added to dominate the milk autoch-thonous microbiota and to favor milk hygiene. The milk is renneted using liquid animal rennet at 32–40 °C for 15–20 min. The coagulum is broken in two steps: the first breakage reduces the curd particles to the size of a walnut and after resting for

Fig. 6.61 Toma Piemonetese. The photo has been kindly supplied by AFIDOP, Italy

a few minutes, the size of the particles is reduced to that of maize or rice. Traditionally, the curd is not cooked but a low cooking is permitted. The broken curd may be held under the whey for 10 min, and then placed into the molds and pressed for up to ca. 5 h to accelerate whey drainage. The curd is maintained at 8–10 °C until the next morning, after which is salted using dry salt for 10–2 days or in saturated brine (18–22°Bé) for 12–48 h. The cheese is ripened for at least 45 days at 8–12 °C and a relative humidity of 80–85%. The final fat content on dry matter must be at least 32%.

The cheese has a cylindrical shape, ca. 7–9 cm high, 30–40 cm in diameter, and weighing 6 to 8 kg.

The cheese paste is elastic with widespread little holes. The cheese rind appears regular and smooth with a light gray color. The color is ivory white and the flavor is sweet or moderately spicy. During ripening, Soft Bra may be further processed by treating the cheese with vinegar or treating only some pieces of the cheese with olive oil.

6.2.18.2 Hard Bra

The manufacture of hard or semi-hard Bra cheese is from whole cow's milk (sometimes supplemented by small quantities of ewe or goat's milk) or sometimes partially skimmed, without curd cooking.

Milk may be refrigerated at 4 °C at the dairy. Traditionally, cheese was produced without starters, but today, natural milk or whey cultures or commercial/selected starters may be used. The milk is coagulated using liquid animal rennet at 27–30 °C for 15–20 min. Also for hard Bra cheese the coagulum is broken in two steps: the

Fig. 6.62 Geographical area for the manufacture of Toma Piemontese cheese

Fig. 6.63 Fromadzo Valle d'Aosta cheese. The photo has been kindly supplied by ONAF (Italian Cheese Taster Organisation, www.onaf.it)

Fig. 6.64 Geographical area for the manufacture of Fromadzo Valle d'Aosta cheese

first breakage reduces the curd particles to the size of beet root and after resting for a few minutes, the size of the curd particles is reduced to that of maize or rice. Traditionally, curd cooking is not permitted. The broken curd may be held under the whey for 10 min, and then placed into the molds and pressed for up to ca. 5 h to accelerate whey drainage. The curd is maintained at 8–10 °C until the next morning, after which it is salted using dry salt for 10–12 days or in brine (18–22°Bé) for 12–48. The final fat content of the cheese dry matter must be at least 32%. The cheese is ripened for at least 120 days at 12–15 °C.

The cheese has a cylindrical shape, ca. 7–9 cm high, 30–40 cm in diameter, and weighing 6 to 8 kg. The texture is hard or semi-hard. The cheese paste is harder than the soft type and may contain a few little holes. The cheese rind appears harder than the soft type, with a dark brown color. The color is yellow-brown and flavor is moderately spicy.

Fig. 6.65 Valtellina Casera cheese. The photo has been kindly supplied by AFIDOP, Italy

6.2.19 Caciotta

Caciotta is a traditional cow's milk cheese, produced mainly in regions in the center of Italy (Liguria, Emilia–Romagna, Umbria, Marche, and Tuscany regions). With the name Caciotta very different cheese types could be produced, depending on the tradition of the different geographical zone, each of them with different organoleptic properties. The name derives from the Italian term *cacio*, which is the vulgar term to indicate the cheese (Mucchetti and Neviani 2006) (Figs. 6.69, 6.70, 6.71, and 6.72).

Usually, Caciotta is intended as a semi-soft cheese with a short–medium ripening time, produced from pasteurized whole cow's milk. Refrigerated cow's milk alone or a mixture of cow's milk and ewe's milk is used for cheesemaking. After pasteurization (71 °C for 15 s), the milk is cooled at 37 °C and usually inoculated with commercial/selected thermophilic and/or mesophilic lactic acid bacteria. After 30 min at 37 °C, liquid calf rennet is added and coagulation occurs within 30 min. Curd breaking is to variable dimensions depending on the type of cheese, but usually the size is not less than a maize grain. After whey drainage and molding, the curd is stored for approximately 4 h at room temperature to favor the lactic acid fermentation. Salting is done using dry salt and further storage occurs at room temperature for 12 h. Ripening at 10 °C lasts approximately 15 days for the fresh type or more often for 2 months for the aged type (dry matter approximately 60%) (Di Cagno et al. 2011).

The cheese is cylindrical in shape, 4–8 cm high and 8–16 cm in diameter, and weighs 0.8–2.0 kg. The cheese paste has an elastic, compact, and soft structure, which becomes less soft during ripening. A few heterogeneously distributed holes

Fig. 6.66 Geographical area for the manufacture of Valtellina Casera cheese

may be present. The color of the cheese paste is light yellow. The rind of the aged variety is thin and yellow. The flavor is delicate, slightly acidic.

6.2.20 Casatella Trevigiana

Casatella Trevigiana is a PDO cheese, extremely soft and produced from whole cow's milk in the Treviso territories (Veneto region) (Mipaaf 2017; AFIDOP 2016) (Figs. 6.73 and 6.74).

The whole milk must have a minimum fat content of 3.2%. Milk is collected and stored at 4 °C, pasteurized and then maintained at 34–40 °C. Natural milk culture starters, consisting of thermophilic lactic acid bacteria, are used to acidify the milk.

Fig. 6.67 Bra cheese. The photo has been kindly supplied by AFIDOP, Italy

This acidifying activity markedly influences the final texture of the cheese. The milk is renneted using liquid or powdered rennet at 34–40 °C for a period ranging between 15 and 40 min. The coagulum is broken in two steps. The rupture of the curd is very mild for getting the final extremely soft texture of the cheese. Between the two breaking steps, the curd is maintained under the whey for 45–55 min. At the end of curd breaking, the curd grains are approximately the size of a hazel nut. The curd stays in the molds for ca. 24–48 h, when whey drainage occurs. The cheese is molded and maintained for ca. 3 h at 25–40 °C. Salting is carried out using brine (18–20°Bé) or dry salt or even adding salt to the vat milk. The cheese is produced in two sizes, piccolo (little) and grande (big) and ripened for 4 to 8 days at 2–8 °C.

The size of the cheese may be small or large and have a variable weight, 0.20–0.70 kg, while the larger cheese weighs 1.8 and 2.2 kg. The cheese has a cylindrical shape, with a diameter of 5–12 and 18–22 cm, and a height of 4–6 cm or 5–8 for the little and big sizes, respectively. Cheese paste is creamy, with a soft and mild texture. Small but not widely distributed holes may be present. Moreover, the texture of the pasta is such as to render the cheese not classified as spreadable. The color of the cheese ranges from white (or porcelain) to cream white. The rind is absent or just perceptible. The flavor and taste are sweet, characteristic of milk, with a slightly acid note. The fragrance is delicate. Considering the fragility of the cheese structure, it can be commercialized only after packaging.

Fig. 6.68 Geographical area for the manufacture of Bra cheese

6.2.21   *Casolet dell'Adamello*

Casolet dell'Adamello is a traditional cheese, produced from partially skimmed raw cow's milk, with a very short ripening time. Traditionally, the cheese is made in Camonica and Palot Valleys, and Eastern Sebino area (Brescia province in Lombary region) (Mipaaf 2017; AFIDOP 2016) (Figs. 6.75 and 6.76).

The raw milk may be cooled to 4 °C, and partially skimmed to ca. 2.8% fat content. Then, the milk is thermized or pasteurized and natural milk or whey culture or commercial/selected thermophilic starters may be added to the vat milk. The milk is coagulated using calf rennet at 36–38 °C. The curd is broken into hazelnut-sized pieces and heated to ca. 40 °C for 5–10 min. Molded curd is held at ca. 35 °C for few hours. During this period, the curd is turned frequently (every 2 h). Salting is

Fig. 6.69 Caciotta cheese type. The photo has been kindly supplied by Prof. Germano Mucchetti (Parma University, Italy)

Fig. 6.70 Caciotta cheese type. The photo has been kindly supplied by Prof. Germano Mucchetti (Parma University, Italy)

carried out by immersion in brine (18–22°Bé) at 15 °C for 6–7 h. Ripening is at 7–9 °C for 5–30 days.

The cheese has a triangular, or eventually square, shape with a slightly wrinkled crust, white-yellow in color. The weight varies from 0.8 to 2 kg. The cheese paste is ivory in color, uniform in consistency, with some 3 mm holes. The flat surfaces are 15–25 cm long and 4–6 cm height. The cheese rind is thin and not smooth, the color is white or light yellow, and frequently with the presence white-gray mold. The cheese paste is soft and compact, usually small holes are present, and the taste is sweet and delicate.

Fig. 6.71   Caciotta cheese type. The photo has been kindly supplied by Dr. Elena Bancalari (Parma University, Italy)

Fig. 6.72   Caciotta cheese type. The photo has been kindly supplied by Dr. Elena Bancalari (Parma University, Italy)

6.2.22   *Castelmagno*

The manufacture of Castelmagno PDO cheese is in the Grana Valley, near Cuneo territories (Piedmont Region) (Mipaaf 2017; AFIDOP 2016; Bertolino et al. 2011; Manzi et al. 2007; Mucchetti and Neviani 2006; Gobbetti 2004; Gobbetti and Di Cagno 2002) (Figs. 6.77 and 6.78). It is a semidry cheese made from cow's milk and rarely small amounts of ewe's milk and/or goat's milk, from a minimum of 5% to a maximum of 20%, may be added. Castelmagno cheese may be considered as a hard blue cheese variety with a compact but friable texture and a moderately piquant flavor.

Fig. 6.73 Casatella Trevigiana cheese. The photo has been kindly supplied by AFIDOP, Italy

Raw cow's milk is partially skimmed, according to a creaming protocol similar to that described for Parmigiano Reggiano and Grana Padano cheeses. The traditional technology does not involve the use of a starter, and acidification is due to the milk autochthonous lactic acid bacteria. Liquid or powdered calf rennet, which may be combined with a small amount of lamb rennet paste, is used for coagulation, usually at 35–38 °C in 30–40 min, After renneting, the coagulum is broken into particles dimension of a walnut or hazel nut. After cutting the coagulum and removal of most of the whey, the curd is traditionally harvested in cloth bags, which are hung for 10–12 h at room temperature, allowing the further removal of whey. Then, a typical technological step is used. The curd compacted in the cloth bags is removed and cut into pieces, and then is transferred into closed containers (plastic or steel) where it is stored under whey at room temperature for 2–4 days, depending on the season (shorter during summer). The whey temperature is maintained constant, at least at 10 °C, and the whey may be partially replaced by adding hotter fresh whey. This stage of the process is probably unique and characteristic of this cheese, but an explanation of its consequences on the cheese has not been given in the literature. Then, the curd is mechanically and manually remixed, compressed for 10–15 min, and molded. During this phase, salt may be added. The curd is pressed for 1–3 days by overlaying stones weighing 5–6 kg, equivalent to the weight of the cheese, or with a mechanical press in order to complete whey drainage. Salting of the cheese is carried out by using dry salt for 20–30 days. Ripening takes place in natural caves at 10–12 °C and 85–90% relative humidity. The cheese microbiota may be extremely variable because of the absence of strongly selective technological stages in the dairy process. When present, the mold mycelium is due to the growth of molds naturally present in the production environment that contaminate cheese. Ripening lasts

Fig. 6.74 Geographical area for the manufacture of Casatella Trevigiana cheese

at least 2 months for the fresh product, and 5 months for the medium-ripened product.

The cheese has a cylindrical shape, ca. 12–20 cm high, 15–25 cm in diameter, and weighs 2–7 kg. *Penicillium* spp. from the environment may be present on the cheese surface, and occasionally in the interior of the cheese. The color of the cheese paste is pearly white or ivory fresh. The cheese surface assumes ochre yellow color on aging and has bluish-green veins due to mold growth. The taste is delicate, moderately salty for young cheese, while strong and spicy, and sometimes very salty, when long-ripened. Sometimes, the aroma becomes ammoniacal in long-ripened cheeses. The presence of ewe's milk makes the flavor even more pronounced.

Fig. 6.75 Casolet dell'Adamello cheese. The photo has been kindly supplied by Prof. Germano Mucchetti (University Parma, Italy)

Fig. 6.76 Geographical area for the manufacture of Casolet dell'Adamello cheese

6.2.23 *Crescenza*

Crescenza is a very soft cheese traditionally produced in northern Italy, particularly in the Lombardy region (Figs. 6.79, 6.80, 6.81, and 6.82). Although it is not a PDO cheese, in Italy it is well known, widespread and appreciated (Mucchetti and Neviani 2006; Salvadori del Prato 1998; Guerriero et al. 1997; Todesco et al. 1992; Ghitti and Ottogalli 1987; Savini 1945; Besana 1915). In the ancient tradition of the Lombardy region, Stracchino cheese was distinguished from Crescenza based on cheese structure, consistency and creaminess. Both are soft cheeses, but Stracchino has higher consistency and a thin crust that is tougher, while Crescenza is soft, without or with a very thin crust, more susceptible to deformation and collapse (Figs. 6.79, 6.81, and 6.82).

Fig. 6.77 Castelmagno cheese. The photo has been kindly supplied by AFIDOP, Italy

Whole cow's milk is collected and stored at 4–6 °C, usually for not more than 24 h but even more, depending on the frequency of milk collection, and then pasteurized at the dairy. When the cheese is produced industrially in large quantities it is possible to standardize the fat content by centrifugation. Cheesemaking could be done in large vats or, depending on the dairy structure, in multiple tanks. Commercial/ selected thermophilic starters, consisting mainly of *Streptococcus thermophilus* strains, are used. Commercial/selected starters and natural milk cultures may sometimes be used together, in which case the commercial/selected thermophilic starter drives the acidification process in the milk and curd. Natural milk cultures consist mainly of mesophilic lactococci and are used as secondary starters to improve the cheese aroma. The milk is renneted using liquid calf rennet (minimum content of chymosin, 70%) at 38–42 °C for 25–35 min. It is possible to add glucone-delta-lactone (GDL) to the milk to facilitate process standardization. Then, the coagulum is broken and molded (Fig. 6.82). Nowadays, most of these procedures are highly mechanized and automated. Molded cheeses are maintained at an ambient temperature of 25–30 °C for ca. 3–4 h, which is followed immediately by salting in brine (16–18°Bé) at 12–15 °C for 1–2 h, depending on the size of the cheese. Crescenza has a very short maturation period (ca. 7 days) at 4 °C and a relative humidity higher than 90%. During ripening, it is necessary to achieve sufficient drying of the surface to allow it to be wrapped in paper without subsequent separation of serum. The final water content is 57–59%.

It is a square- or rectangular-shaped cheese with a side of 10–20 cm and 5–6 cm high. Crescenza is a very delicate cheese and to obtain the desired structure is a complex technological goal. The final high water content and its soft consistency must be balanced. Firstly, it is necessary to avoid the defect of cheese serum drainage, in Italy called *colatura*. This defect is due to excessive proteolysis due to residual rennet activity in the cheese paste that could cause the loss of structure and

Fig. 6.78 Geographical area for the manufacture of Castelmagno cheese

shape of the cheese. Depending on the intensity of proteolysis, different regional products of different consistency may be obtained (e.g., Stracchino and Squacquerone). At the same time, the excessive acidification caused by starter activity demineralizes caseins excessively and leads to a harder and more friable structure and could produce roughness of the cheese paste. The structure of the cheese paste is homogeneous, very soft and smooth consistency, without rind. The color of the cheese paste is white or slightly straw-yellow. The presence of little holes is possible but is not appreciated. The flavor is mild, slightly acidic, and bitter.

Fig. 6.79 Crescenza cheese. The photo has been kindly supplied by Dr. Giorgio Giraffa (CRA-Mipaaf- Lodi, Italy)

6.2.24 Gorgonzola

Gorgonzola is one of the most important and famous Italian PDO cheeses (Mipaaf 2017; AFIDOP 2016; Bernini et al. 2016; Bertolino et al. 2011; Manzi et al. 2007; Mucchetti and Neviani 2006; Carminati et al. 2004; Moio et al. 2000; Gobbetti et al. 1997a; Contarini and Toppino 1995; Carini 1990; Martelli 1989; Arnaudi 1948; Zucconi 1874). It is an ancient cheese but its exact origin is uncertain because several areas in the Lombardy region claim to be its birthplace. Today, the cheese is produced only in specific areas of Piedmont and Lombardy regions (Bergamo, Brescia, Como, Lecco, Lodi, Cremona, Milano, Pavia, Varese, Monza, Biella, Cuneo, Novara, Vercelli, Verbano-Cusio Ossola, and Casale Monferrato) (Figs. 6.83, 6.84, 6.85, 6.86, 6.87, 6.88, 6.89, 6.90, 6.91, 6.92 and 6.93).

Gorgonzola is a blue-veined, mold-ripened cheese made from pasteurized cow's milk. Two principal types of Gorgonzola are produced: Gorgonzola Dolce (Sweet Gorgonzola), sweet and creamy, and Gorgonzola Piccante (Piquant Gorgonzola) with a stronger flavor and more compact cheese paste structure. Both cheeses have the typical blue-veined network in the cheese paste.

The milk used for Gorgonzola comes from cows bred in the territory of cheese production and may be stored at 4–6 °C. At the dairy, the milk is pasteurized and then cultures of lactic acid bacteria and molds are added. Traditional hemispherical vats (Fig. 6.85) or multiple tanks are used. The pasteurized milk is inoculated with natural milk cultures or commercial/selected thermophilic starters containing mainly *Lactobacillus delbrueckii* subsp. *bulgaricus* and *Streptococcus thermophi-*

Fig. 6.80 Geographical area for the manufacture of Crescenza cheese

lus, together with *Penicillium roqueforti* that develops as an internal blue-green mold. In order to favor mold growth it is useful to have a cheese paste structure permeable to oxygen. To obtain this goal, a special attention must be payed to curd breaking and molding. Auxiliary starters of mesophilic lactic acid bacteria and/or yeasts may also be used. The synthesis of CO_2 promotes openness of the cheese paste (in Italian *apertura della pasta*). Further veining is a result of the growth of edible mold, a process called *erborinatura* in Italian, which develops as the cheese ripens and causes its typical flavor, taste and consistency. The milk is coagulated using liquid calf rennet at 28–36 °C for 25–30 min. The coagulum is broken into large pieces (cubes) with a side of about 5 cm, which are collected into molds and placed on drainage tables (Figs. 6.85, 6.86, and 6.87). The curd may be cooled at 8–10 °C for 12–15 h before being transferred to warm rooms in which salting with dry NaCl is performed during few days (Fig. 6.88). Ripening takes place in rooms

Fig. 6.81 Crescenza cheese. The photo has been kindly supplied by Dr. Elena Bancalari (University Parma, Italy)

Fig. 6.82 Crescenza cheesemaking. Curd molding. The photo has been kindly supplied by Prof Germano Mucchetti (University Parma, Italy)

the temperature and humidity of which vary during the maturation process: initially at 2–4 °C and 95% humidity (Fig. 6.89). After ca. 15 and 25 days, the curd is pierced to facilitate aeration of the paste necessary for the development of *P. roqueforti*, characteristic of Gorgonzola, and determining the typical green appearance of the paste (Fig. 6.90). During this phase, the structure, flavor and aroma undergo radical changes. *P. roqueforti* catabolizes the lactic acid produced by starter lactic acid bacteria and neutralize the pH of the curd. Intense proteolysis and lipolysis start, with the formation of aromatic compounds that ensure the unique sensory profile of the cheese. Gorgonzola Dolce has a minimum ripening period of 50 days and a

Fig. 6.83 Gorgonzola cheese. The photo has been kindly supplied by AFIDOP, Italy

maximum of 150 days (Figs. 6.91 and 6.92). It develops a soft creamy consistency, an unmistakable flavor, and sweet taste. Gorgonzola Piccante has more blue veining in the cheese, and is firmer and slightly crumbly. This cheese has a minimum ripening period of 80 days and a maximum of 270 days. During ripening, Piccante type develops a very spicy taste and a strong flavor. The other type Gorgonzola Piccolo Piccante (Small Piquant Gorgonzola) has deep blue veining in the cheese, and is firmer and slightly crumbly (Fig. 6.93). This type has a minimum ripening period of 70 days and a maximum of 200 days; its taste and flavor are very similar to Gorgonzola Piccante. The old technology of the Gorgonzola *due paste* (two paste) was based on a very different concept of production, which has been practiced for decades. The *due paste* technology consisted of producing two distinct curds from the milk derived from each of two milking. The evening cooled curd was combined with the warm curd just made in the morning by alternating layers of the two curds and taking care to leave on top and bottom the morning curd layers. However, it is not possible to perfectly combine the two different curds and within the cheese small interstices remain in which, during ripening, the green mold develops. The reasons for the abandonment of this technology were primarily of an industrial nature, in the light of knowledge on hygiene and health issues.

The shape of the cheese is cylindrical, with a diameter of 20–32 cm and a minimum height of 13 cm. As described above, three types are coded: Piccante, Piccolo Piccante, and Dolce. The weight ranges from 9 to 3.5 kg for the Dolce and Piccante types and between 5.5 and 9.0 kg for the Piccolo Piccante type. The duration of ripening determines whether the product has a sweet, mild taste or a slightly sharp taste, after an aging period of at least 50 days, or a clearly sharp taste after a ripening period of at least 70 or 80 days, depending also on the weight of the wheel. The cheese paste is characterized by the presence of numerous green-blue veins due to the growth of the molds in the holes left by the needles. There are no holes in paste

Fig. 6.84 Geographical area for the manufacture of Gorgonzola cheese

that is creamy and mild in the Dolce type, more compact in the Piccante type. The cheese has a rough rind, colored from gray to rose, depending on the prevailing microbiota. On one face is embossed the consortile logo. The rind is declared not edible (Gazzetta Ufficiale Italiana—Italian Official Journal 138 of 16/6/2001). Flavor and taste are characteristics. Cheeses are present in the market after portioning.

Fig. 6.85 Gorgonzola cheesemaking. Curd breaking. The photo has been kindly supplied by AFIDOP, Italy

6.2.25   Graukase

Graukase is a cow's milk cheese belonging to the *sauerkäse* variety, which is widespread on the Alp territories of the Tyrol (Figs. 6.94 and 6.95). Literally, Graukase a translation from the German means gray cheese due to the abundant growth of gray molds. Its main distinguishing feature is the acid coagulation without the use of renneting. The main region of manufacture is in Bolzano province (Trentino Alto Adige region) in typical farms called *Maso* during the period between June and September. This period coincides with pastures on the Alps.

Raw cow's milk from a single milking is fully or partially skimmed and warmed to ca. 25 °C. The practice of skimming derives from the old tradition of villages where it has been important to recover the fat from milk for nutritional purposes and

Fig. 6.86 Gorgonzola cheesemaking. Curd molding. The photo has been kindly supplied by AFIDOP, Italy

where rennet was absent. Natural milk or whey cultures may be used to inoculate the skimmed milk. After ca. 24–36 h, acid coagulation occurs and the curd is placed on linen cloths where it is subjected to some pressure. Then, the curd is broken manually into large pieces, dry salted, and pepper added. A second breaking takes place to the dimensions of maize grains and during this operation pieces from a previous ripened cheese are mixed. After this mixing, the mixture of curd and cheese pieces is molded, ripened at room temperature and under humid conditions for ca. 10 days.

The cheese has cylindrical or parallelepiped shape, irregular or very irregular, weighing approximately 1–1.5 kg. The cheese paste is crunchy, granular, soft, white or straw-colored. There are mechanical openings that make the paste very uneven, and there is no rind. The flavor and taste are milky, acidic with some spicy notes.

6.2.26   Italico

As the name attests, Italico is a very traditional cheese produced in the Lombardy region (particularly in Pavia territory), even if it is not recognized as a PDO cheese (Mucchetti and Neviani 2006; Ghitti and Bianchi-Salvadori, 1985) (Figs. 6.96 and 6.97). The best-known brand product is Belpaese. Strangely, the brand Belpaese is better known and recognized than the general name Italico. It is a semi-soft cheese with a short–medium ripening time, produced from pasteurized whole cow's milk.

The whole milk is collected and stored at 4–6 °C and then pasteurized at the dairy. Usually, the cheese is industrially produced in large quantities and it is possible to standardize the fat content by centrifugation. Cheesemaking could be done in multiple tanks. Natural milk cultures or commercial starters, consisting of ther-

Fig. 6.87 Gorgonzola cheesemaking. Curd molding. The photo has been kindly supplied by AFIDOP, Italy

mophilic lactic acid bacteria such as *Streptococcus thermophilus*, are used. The milk is renneted using 40–55 ml liquid calf rennet per 100 kg at 40–43 °C for 20–25 min. Then, the coagulum is broken into medium-large pieces and molded. Most of these procedures are highly mechanized and automated. Whey drainage takes place in a tunnel or in warm rooms at 25–30 °C and relative humidity 90–95% for 5–8 h. The pH of the curd at the end is ca. 5.0. Salting is in brine (18°Bé) at 10–15 °C for 7–12 h. Ripening for ca. 30–40 days (even 60 days) is in a climate-controlled room at 4 °C and a relative humidity higher than 90%.

The cheese has a cylindrical shape, 6–8 cm high, 10–20 cm in diameter, and a final weight of 1.6–2.2 kg. The cheese paste is elastic, without eyes. The color is white to slightly straw yellow. The rind is thin, smooth, usually straw-yellow in color, but sometimes slightly pinkish. The characteristic flavor is sweet and buttery.

Fig. 6.88 Gorgonzola cheesemaking. Curd dry salting. The photo has been kindly supplied by the Consorzio Gorgonzola PDO, Italy

6.2.27   Pannerone

Pannerone is a traditional cheese produced in the territory of Lodi (Lombardy region) (Mucchetti and Neviani 2006; Tripi 1980; Delforno 1982; Ottogalli et al. 1975; Savini 1950) (Figs. 6.98 and 6.99). This cheese has one traditional and very specific feature, the total absence of any salting process. The name Pannerone derives from *panéra*, which in the dialect of Lodigiano means cream of milk. At one time it was also named White Gorgonzola, although it did not really have any kinship with Gorgonzola, except for some exterior appearances and sometimes its weight.

Fig. 6.89 Gorgonzola cheesemaking. Early curd ripening. The photo has been kindly supplied by the Consorzio Gorgonzola PDO, Italy

Fig. 6.90 Gorgonzola cheesemaking. Curd piercing. The photo has been kindly supplied by the Consorzio Gorgonzola PDO, Italy

This fresh cheese is obtained from raw whole cow's milk, without starter, without curd cooking, without salting and with a short ripening. Raw milk is not refrigerated and not heat-treated. Renneting is done with liquid calf rennet at 28–32 °C in ca. 30 min. The curd is broken until pieces the size of a bean are reached. The curd is recovered and manually crumbled and molded. The molded curd is held for 4–8 days at 28–32 °C, to complete whey drainage. Ripening continues by keeping the cheese at room temperature for 24–48 h, and then finally in a refrigerated room at 4–6 °C. Overall, the curing time is ca. 10 days. The cheese is ready for consumption

Fig. 6.91 Gorgonzola cheesemaking. Curd ripening. The photo has been kindly supplied by the Consorzio Gorgonzola PDO, Italy

Fig. 6.92 Gorgonzola cheesemaking. Cheese assessing. The photo has been kindly supplied by the Consorzio Gorgonzola PDO, Italy

Fig. 6.93 Gorgonzola cheesemaking. Cheese cutting. The photo has been kindly supplied by the Consorzio Gorgonzola PDO, Italy

Fig. 6.94 Graukase cheese. The photo has been kindly supplied by the Federazione Latterie Alto Adige, Italy

after about 15–20 days and may be ripened for up to 2 months. Cheese microbiota relies on milk autochthonous species and different bacteria and yeasts are involved in the alcoholic and heterolactic fermentations. *Enterobacteriaceae, Enterococcus*, lactic acid bacteria, *Staphylococcus* and yeasts may be present. Of course the technology of this particular and old cheese, produced by a few small dairies, is debated mainly concerning the hygienic aspects. It is even necessary to underline that this is a very limited production, and that to date no poisoning or infections resulting from the consumption of Pannerone have been reported.

The cheese has a cylindrical shape, with a diameter of 20–30 cm, a height of ca. 20 cm and weighing 10–13 kg. The cheese paste is soft and fragrant, is white in

Fig. 6.95 Geographical area for the manufacture of Graukase cheese

color and shows a large number of holes. The rind is thin and smooth, elastic and with a color ranging from yellow to pink. The characteristic flavor and taste are sweet and bitter at the same time, but non-acid.

6.2.28   Quartirolo Lombardo

Quartirolo Lombardo is a PDO soft cheese. Quartirolo is produced in a specific geographical area of Lombardy region, particularly in the territory of Bergamo, Brescia, Como, Cremona, Lecco, Lodi, Monza, Milano, Pavia, Varese (Mipaaf

Fig. 6.96 Italico cheese

2017; AFIDOP 2016; Manzi et al. 2007; Mucchetti and Neviani 2006; Spinelli 1958) (Figs. 6.100 and 6.101).

Quartirolo Lombardo is produced from whole pasteurized cows' milk obtained at two milking, in some cases with the addition of some partially skimmed milk. Usually, the milk from the first milking must be used whole; the milk from the second or subsequent milkings may be used whole or partially skimmed.

The whole milk is collected and stored at 4–6 °C. Even if the production of raw milk cheese is permitted, today, milk is generally pasteurized at the dairy. Usually, a commercial/selected starter composed mainly of *Streptococcus thermophilus* and *Lactobacillus delbrueckii* subsp. *bulgaricus* strains is used. The milk is renneted using liquid calf rennet at 35–40 °C for 20–25 min. The coagulum is broken, usually in two steps with a break of ca. 10 min. The curd grains obtained at the end of breaking are the size of a hazelnut and are then molded. As in the case of Taleggio cheese, most of the cheesemakers nowadays are highly mechanized and automated. Whey drainage occurs in air-conditioned rooms at 26–28 °C for 4–24 h. Under such conditions the thermophilic lactic acid bacteria develop. Traditionally, salting is accomplished by using brine or dry salt over a few days. Ripening is usually for 30 days, but also for a few months, in rooms at a temperature between 2 and 8 °C and a relative humidity of 80–95%. Cheese ripening, along with the related biochemical events, proceeds more intensively from the rind to the center of the cheese, as result of the surface contamination by the environmental microbiota.

The cheese has a square-shaped parallelepiped shape, with flat sides and a straight heel, with a length of 18–22 cm and 4–8 cm high. The final weight of the cheese is ca. 1.5–3.5 kg. The structure of the cheese paste is compact, slightly grainy with possible flakes, crumbly (without a yellowish layer under the rind), and becomes firmer, soft and melting during cheese ripening. The color of the paste ranges from white to yellowish white, which may become more intense, depending on the degree of ripening. The cheese rind is thin, soft and pinkish white for the

Fig. 6.97 Geographical area for the manufacture of Italico cheese

young (fresh) cheese and reddish gray-green for the ripened cheese, striped, embossed on one face with the consortile logo, affixed during molding. The flavor is slightly acidic and aromatic when fresh, particularly if the cheese is produced from partially skimmed milk, more aromatic when produced with whole milk and ripened longer.

6.2.29 Raschera

Raschera is a PDO cheese produced in Cuneo territories (Piedmont region) (Mipaaf 2017; AFIDOP 2016; Manzi et al. 2007; Mucchetti and Neviani 2006; Pavolotti 1981) (Figs. 6.102 and 6.103). It is a semi-soft or semihard cheese, depending on the

Fig. 6.98 Pannerone cheese. The photo has been kindly supplied by Prof. Germano Mucchetti (Parma University, Italy)

type. It is a fatty or semi-fat cheese, produced from cow's milk, possibly with small additions of ewe and/or goat's milk. In any case, it is necessary to have a minimum fat content of 32% in the ripened cheese. If the manufacture takes place on farms located above 900 m high, the name of the variety is Raschera d'Alpeggio cheese.

The manufacture is from raw, refrigerated, thermized, or pasteurized milk. The milk is renneted using liquid animal rennet at 27–30 °C for 10–60 min; sometimes, supplemented with a small amount of rennet paste occurs. Curd breaking and molding are somewhat similar to those used for Bra cheese. Salting is carried out using dry salt and ripening is at 12–15 °C. The minimum ripening period is 30 days for cheese produced from pasteurized milk, and 60 days for that from raw cow's milk.

The cheese may have different shapes and sizes: a cylindrical shape, ca. 6–9 cm high and 30–40 cm in diameter, and weighing 5–8 kg, or a square-shaped parallel-epiped shape, with a length of 28–40 cm and a height of 7–15 cm, and weigh 5–9 kg. The cheese paste has a white or ivory white color, is rather consistent, elastic, and flexible with sparse and irregular holes. For Raschera d'Alpeggio, the color of the cheese paste tends to an intense yellow. The rind may be thin, elastic, smooth and regular, of reddish gray sometimes with yellowish shades, which become more accentuated with ripening. The flavor is fine, delicate, typically fragrant and moderately spicy and flavorful if long-ripened. For the Raschera d'Alpeggio variety, aromas due to fresh fodder may be present.

Fig. 6.99 Geographical area for the manufacture of Pannerone cheese

6.2.30   *Robiola di Roccaverano*

Robiola di Roccaverano cheese is a PDO soft unripened cheese produced in a specific geographical area of Piemonte region, and in the territory Alessandria and Asti (Mipaaf 2017; Mucchetti and Neviani 2006; Grassi et al. 2002; Pattono et al. 2001; Coisson et al. 2000) (Figs. 6.104 and 6.105). Traditionally, it is produced from goat's milk, but it is permitted to add cow's milk, to a maximum of 85%, and sheep's milk. It is produced principally by lactic acid fermentation and usually without ripening.

The whole milk is collected and stored at 20–25 °C. Under these conditions, milk autochthonous mesophilic lactic acid bacteria multiply. Milk could be partially skimmed but at the end of production, the cheese must have a minimum fat content of 45% in the dry matter. The milk is renneted using liquid rennet at 18–28 °C for 12–36 min. The coagulum is not broken or heat treated. Curd is placed directly into

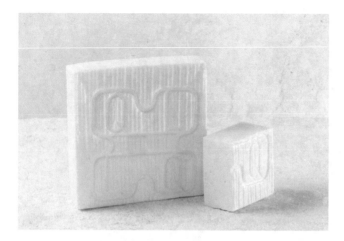

Fig. 6.100 Quartirolo Lombardo cheese. The photo has been kindly supplied by AFIDOP, Italy

the mold as described by the French method called *moulage à la louche*. The curd remains in the mold for ca. 24–48 h and under these conditions whey drainage occurs. Traditionally, the cheese is salted using fine dry salt. If the cheese is held for some days, cooling is necessary.

The shape of the cheese is cylindrical, 4–5 cm high and 10–14 cm in diameter, with a weight of 250–400 g. The structure of the cheese paste is soft, tending to become compact after longer conservation times. Cheese paste is white with a finely granular texture, having the typical features of those varieties made with a marked contribution of lactic acid fermentation, with a tendency to assume a more creamy texture when the cheese is aged for 2–3 weeks. Usually, the cheese is without a rind, but if it held for some weeks, a thin, hard-rind forms. The flavor is delicate, slightly acidic, with the typical flavor of goat's milk that can become strongly flavored and spicy on longer storage.

6.2.31   Salva Cremasco

Salva Cremasco is a DOP soft smear-ripened cheese or washed-rind cheese, produced exclusively with whole cow's milk, with a minimum of 75 days of ripening. The production area includes the entire territory of the provinces of Bergamo, Brescia, Cremona, Lecco, Lodi, and Milan (Lombardy region), where all the operations of milk production, dairy, and ripening must take place (Mipaaf 2017; AFIDOP 2016; Mucchetti and Neviani 2006) (Figs. 6.106 and 6.107).

Milk may be pasteurized, and a natural milk or whey culture or a commercial/selected starter may be added to the vat milk. Liquid bovine rennet is used to coagulate the milk at 32–40 °C in 10–20 min. The coagulum is broken in different steps to yield curd particles the size of a hazelnut. The curds are not heated. Molded

Fig. 6.101 Geographical area for the manufacture of Quartirolo cheese

cheese can be held for a minimum of 8 h to a maximum of 16 h at a temperature of 21–29 °C and a relative humidity of 80–90%. The curd is salted with dry salt or in brine (16–18°Bé). The curd is ripened at 2–8 °C for at least for 75 days. During ripening, the cheese is turned frequently and treated with brine or dry salt. No treatment of the crust is allowed, except by sponging with water and salt.

The cheese looks like a parallelepiped with quadrangular shape and a variable weight from 1.3 to 5 kg. The cheese paste has few irregularly distributed holes and a white color tending to strawberry with increasing ripening time. It is compact and friable but softer in the part immediately below the rind due to maturation, which proceeds from the exterior to the interior of the cheese. The cheese rind is thin, smooth, of medium consistency with presence of characteristic biota. The flavor of the cheese is aromatic and intense, being more pronounced with aging.

Fig. 6.102 Raschera cheese. The photo has been kindly supplied by AFIDOP, Italy

6.2.32 Squacquerone di Romagna

Squacquerone di Romagna is a PDO very soft cow's milk cheese, with very short ripening (Mipaaf 2017). The area of manufacture includes the provinces of Ravenna, Forli, Cesena, Rimini, and Bologna, and a part of the territory of the province of Ferrara in the Emilia–Romagna region (Figs. 6.108 and 6.109).

The whole milk used must have fat and protein contents not less than 3.5% and 3% (w/v), respectively. Milk must be collected within 48 h from the first milking. The milk delivered to dairies must have a temperature less than 10 °C, and storage is at less than 6 °C. Before cheesemaking, raw milk is pasteurized or thermized. Natural milk or whey cultures, composed mainly of *Streptococcus thermophilus* strains, are added to the vat milk. The milk is renneted at 35–40 °C with liquid calf rennet, in 10–30 min. After renneting, the curd is broken into particles of walnut size. The curd must retain an elevated level of water to give the typical creaminess and spreadability. To favor cheese acidification, the broken curd is left to rest under whey for at least 5 min at 35–40 °C and then stirred. The pH reaches 5.9–6.2. The curd is molded and turned at least once in 24 h to favor whey drainage. The molded curd is left at room temperature for a maximum of 3 h, after which it is placed at a temperature not exceeding 15 °C. Salting is in brine (16–18°Bé) at a temperature below 20 °C for 10–40 min, depending on the size. The addition of salt to the milk is permitted also, 400–800 g per 100 L of milk, but only before renneting. Ripening lasts for 1–4 days at ca. 3–6 °C. The moisture content of the cheese varies between 58 and 65%, and the fat content is 46–55% in cheese dry matter. The cheese has a pH between 4.9 and 5.3.

The shape of the cheese depends on the size and form of the box used for molding. The weight ranges from 0.1 to 2.0 kg. The cheese paste is pearly white. The texture is soft, very creamy and speadable, without a compact structure; a rind is absent. The taste is sweet and a little acidic. The flavor is that typical of milk, sometime with an herbaceous note.

Fig. 6.103 Geographical area for the manufacture of Raschera cheese

Fig. 6.104 Robiola Roccaverano cheese. The photo has been kindly supplied by ONAF (Italian Cheese Taster Organisation, www.onaf.it)

Fig. 6.105 Geographical area for the manufacture of Robiola Roccaverano cheese

6.2.33   *Taleggio*

Taleggio is PDO medium-soft smear-ripened cheese, produced in specific geographical areas of the Lombardy and Pidemond regions, particularly in the provinces of Bergamo, Brescia, Como, Cremona, Milano, Pavia, Novara, and Treviso (Mipaaf 2017; AFIDOP 2016; Manzi et al. 2007; Mucchetti and Neviani 2006; Cantoni et al. 2005; Corsetti et al. 2001; Gobbetti et al. 1997b; Ottogalli et al. 1996; Delforno 1967; Carini et al. 1969; Spinelli 1958; Renko 1956) (Figs. 6.110, 6.111, 6.112, 6.113, 6.114, 6.115, 6.116, 6.117, 6.118, 6.119 and 6.120). The cheese is manufactured from whole cow's milk stored at 4–6 °C and pasteurized at the dairy (Fig. 6.112). Nowadays, raw milk is used. Usually, commercial/selected starters, consisting mainly of *Streptococcus thermophilus* and *Lactobacillus delbrueckii* subsp. *bulgaricus* strains, are used. The milk is renneted using liquid calf rennet at

Fig. 6.106 Salva Cremasco cheese. The photo has been kindly supplied by AFIDOP, Italy

34–37 °C for 20–30 min. The coagulum is broken in two steeps with a break of ca. 10 min (Figs. 6.113 and 6.114). The curd grains at the end of breaking are about the size of a hazelnut and are then molded (Figs. 6.115, 6.116, and 6.117). Most cheesemakers may be highly mechanized and automated. Whey drainage occurs in air-conditioned rooms at 22–25 °C for 8–16 h, during which the thermophilic starters develop. Traditionally, the curd is salted using dry salt, which is added over a few days. Nowadays, also brining could be used (Figs. 6.118 and 6.119). Ripening lasts at least 35 days at 2–6 °C at a relative humidity of 85–90% for a few months. The extent of ripening and the consequent biochemical events proceed from the surface to the center of the cheese. Following this trend, proteolysis and lipolysis cause the characteristic creamy aspect of the layer under the rind, while the crust becomes slightly colored (Fig. 6.120). The cheese could be defined also as a smear-ripened or washed-rind cheese (in Italian *formaggio crosta lavata*). The surface microbiota is selected during ripening by washing and brushing the crust with brine or saline solution and it, therefore, consists of aerobic and salt-tolerant species. A heterogeneous microbiota populates the surface, mainly consisting of eumycetes (e.g., *Geotrichum candidum, Penicillium* spp., *Debariomyces hansenii*) and bacteria (e.g., *Micrococcaceae, Microbacterium* spp.*, Brevibacterium* spp.*, and Arthrobacter*), which are responsible for the pink coloration of the rind and of the classical ripening (Ottogalli et al. 1996). Unlike other washed-rind cheeses, the washing is done to ensure uniform growth of desired bacteria or fungi and to prevent the growth of undesirable molds. The increase in pH, together with the above surface treatment, is one of the causes that make it possible to spread the surface contamination of the crust. The thermophilic lactic acid bacteria starters prevail in the cheese paste.

The cheese has a square-shaped parallelepiped shape, with a length of 18–20 cm and a high 5–8 cm, with a final weight of 1.7–2.2 kg. The structure of the cheese paste varies from crumbly to creamy, depending on the distance from the crust and

Fig. 6.107 Geographical area for the manufacture of Salva Cremasco cheese

the degree of ripening. It has a uniform and compact structure without eyes. The rind is thin and pink, embossed on one face with the consortile logo, affixed during molding. The rind microbiota activity causes proteolysis and lipolysis of the cheese paste below the rind, which becomes softer than the central part of the cheese that remains more elastic. The color of the crust, measured by objective methods, is one of the requirements characterizing the product compliance with the identification standard. Taleggio has a distinctive pinkish or orange coloring in the exterior. The flavor of the cheese is slightly aromatic, with no bitter or sharply acidic attributes. The cheese rind is considered edible by the Consortium of Taleggio PDO Cheese, which suggests light scraping before consumption.

Fig. 6.108 Squacquerone di Romagna cheese. The photo has been kindly supplied by ONAF (Italian Cheese Taster Organisation, www.onaf.it)

6.3   Ewe's Milk Cheeses

Although quantitatively less than cow's milk cheeses, the manufacture of ewe's milk cheeses in Italy is consistent (ca. 41,500 tonnes in 2016) and based on a solid tradition (Assolatte 2016; Mucchetti and Neviani 2006). Primarily, most of the ewe's milk cheeses refer to the common name "Pecorino," which derives from the Italian *pecora* (sheep). Apart from those listed in this section, it could be stated that almost each region or province, mainly in the center and south of Italy, has its own Pecorino cheese with names that refer to the place of manufacture (e.g., Pecorino Siciliano, Pecorino Sardo). Most of the ewe's milk cheese varieties are manufactured as table and grated cheeses. Therefore, the types of each cheese may differ markedly with the time of ripening. The use of raw or pasteurized ewe's milk alone or in a mixture with a low proportion of cow's milk, natural milk or whey cultures, rennet paste, low or absent cooking of the curd and intense and moderate to pronounced piquant flavor, especially for long-ripened varieties, are some of the main features that distinguish the Italian ewe's milk cheese production.

Fig. 6.109 Geographical area for the manufacture of Squacquerone di Romagna cheese

6.3.1   *Canestrato di Moliterno*

Canestrato di Moliterno is a PGI (Protected Geographical Indication) cheese manufactured from sheep's or goat's milk. The production is limited to various areas
close to Potenza and Matera cities (Basilicata region) (Mipaaf 2017) (Figs. 6.121
and 6.122). The PGI recognition refers only to cheeses made from a mixture of
whole ewe's and goat's milk at 70–90% and 10–30%, respectively.

   After milking, the milk is refrigerated. Raw milk from one or two consecutive
milkings is used for cheesemaking, before 48 h from milking. Raw or thermized
milk may be used. Natural milk or whey cultures or commercial starters may be
used to inoculate the vat milk. Renneting of milk is using lamb or kid rennet paste,
at a temperature between 36 and 40 °C within 35 min. The curd is broken to obtain
particles the size of rice grains. After a few minutes of resting under whey, the

Fig. 6.110 Taleggio cheese. The photo has been kindly supplied by AFIDOP, Italy

extracted curd grains are molded in the *canestro* made of rushes or other food-grade materials. Pressing by hand occurs to favor the whey drainage. The curd may stay in whey at a temperature not exceeding 90 °C for not more than 3 min, just to get fast cooking and rind formation. The cheese is salted using dry salt or brine. The cheese microbiota depends on milk autochthonous strains (mainly mesophilic lactic acid bacteria and enterococci) and to the thermophilic lactic acid biota from the starters. Based on the period of ripening, the following types are distinguished: (1) primitive, with maturation up to 6 months; (2) seasoned, with maturation over 6 months up to 12 months; and (3) extra, with maturation over 12 months. The ripened cheese has a fat content higher than 30% of the dry cheese matter. The moisture content may not exceed 40%.

The cheese has a cylindrical shape, 10–15 cm high and 15–20 cm in diameter; the weight ranges from 2 to 5.5 kg. Usually, the highest weight is for long-ripened cheese. The cheese paste is hard, sometimes with nonuniformly distributed small holes. The color varies from white to intense yellow, depending on the ripening time. The rind has a pale yellow color, tending to brown or black, if the cheese is treated with oil or a mixture of water and ash. The cheese has a sweet taste, while spicy notes characterize the long-ripened types.

6.3.2   Canestrato Pugliese

Canestrato Pugliese is a PDO cheese manufactured only in the Apulia region. The cheese derives its name and traditional shape from the rush basket, *canestro*, in which the curd is ripened (Mipaaf 2017; Mucchetti and Neviani 2006; Albenzio et al. 2001; Corbo et al. 2001; Santoro and Faccia 1998) (Figs. 6.123 and 6.124).

Fig. 6.111   Geographical area for the manufacture of Taleggio cheese

Raw, whole ewe's milk of one or two daily milkings is generally used, but thermized or pasteurized milk may also be processed. A natural whey culture, composed mainly of thermophilic lactic acid bacteria, may be added, and liquid or
powdered calf rennet, or, exceptionally, lamb rennet paste is used. After cutting the
coagulum to particles of 0.5–1.0 cm in size, the curd grains are heated at 45 °C
under whey for 5–10 min or at 80 °C for 30 s, when using raw ewe's milk. Generally,
this treatment is not equivalent to a cooking process. Molding is in the typical *canestro*, and the curd is held at a very variable temperature, 20–42 °C, for ca. 14 h to
favor further acidification. The curd is dry-salted for ca. 2 days and, during ripening
(4–12 months) in the *canestro*, is turned regularly and rubbed with a mixture of oil
and vinegar. Ripening in the *canestro* is limited to a few days for industrial production. Colonization of the surface by molds from the environment frequently becomes
evident during ripening, which are removed by brushing after few months.

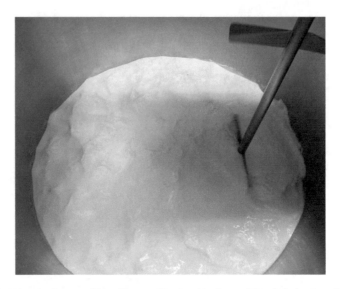

Fig. 6.112 Taleggio cheesemaking. Cheese milk placed in the vat. The photo has been kindly supplied by the Consorzio Taleggio PDO, Italy

Fig. 6.113 Taleggio cheesemaking. Curd breaking. The photo has been kindly supplied by the Consorzio Taleggio PDO, Italy

The cheese has a cylindrical shape, 10–14 cm high and 25–34 cm in diameter, and weighs 7–14 kg. The rind is brown to pale yellow, and the interior cheese paste is compact with small eyes. The flavor is pronounced and tends to be moderately piquant.

Fig. 6.114 Taleggio cheesemaking. Curd breaking. The photo has been kindly supplied by the Consorzio Taleggio PDO, Italy

Fig. 6.115 Taleggio cheesemaking. Curd extraction from the vat before molding. The photo has been kindly supplied by the Consorzio Taleggio PDO, Italy

6.3.3 *Casciotta di Urbino*

Casciotta di Urbino is a PDO semi-cooked cheese produced from a blend of 70–80% whole ewe's milk and 20–30% of cow's milk derived from two daily milkings on farms located in the production area. The cheese is produced throughout the territories of Pesaro and Urbino, in the Marche region (Mipaaf 2017; AFIDOP-IGP 2016;

Fig. 6.116 Taleggio cheesemaking. Curd molding and drainage. The photo has been kindly supplied by the Consorzio Taleggio PDO, Italy

Fig. 6.117 Taleggio cheesemaking. Curd molding. The photo has been kindly supplied by the Consorzio Taleggio PDO, Italy

Manzi et al. 2007; Mucchetti and Neviani 2006; Pinarelli 2005) (Figs. 6.125 and 6.126).

Milk may be refrigerated and pasteurized and inoculated with a natural milk or whey culture or a commercial starter. The mixture of ewe's milk and cow's milk is coagulated at ca. 35 °C for ca. 20–30 minutes with liquid and/or powdered rennet. After molding, the cheese is subjected to manual pressing. Salting is done dry or by alternating brine and dry salting. The cheese is ripened at 10–4 °C and humidity of 80–90%, depending on the size of the cheese, for a period ranging from 20 to

Fig. 6.118 Taleggio cheesemaking. Curd dry salting. The photo has been kindly supplied by the Consorzio Taleggio PDO, Italy

30 days, giving rise to a soft or to a more ripened cheese, respectively. In order to avoid the emergence of molds, the curd may be covered with a transparent and shiny wax to protect the rind. The final fat content must be higher than 45%.

The cheese has a low bowed cylindrical shape with rounded faces. The height is variable between 5 and 7 cm and the diameter is 12–16 cm. The cheese has a variable weight from 0.8 to 1.2 kg, depending on the shape of the mold. Its color is straw-white. The cheese paste is soft and friable with little holes homogeneously distributed but not overly large. The rind is thin, about 1 mm thick, of straw yellow color. The taste and flavor are sweet, characteristic of the particular production procedures.

6.3.4   Fiore Sardo

Fiore Sardo is a PDO hard cheese made from whole raw ewe's milk with medium maturation. The production of Fiore Sardo cheese is strictly limited to some provinces of Sardinia (Mipaaf 2017; AFIDOP-IGP 2016; Mangia et al. 2008; Manzi et al. 2007; Pirisi et al. 2007; Pisano et al. 2006; Mucchetti and Neviani 2006; Ledda et al. 1994; Bottazzi et al. 1978; Ledda et al. 1978; Pettinau et al. 1978; Fascetti 1935) (Figs. 6.127 and 6.128).

Raw whole ewe's milk from a single milking is used. A large part of the milk is produced by an indigenous breed of sheep. The milk may be stored at 4 °C for a maximum of 36 h before cheesemaking. Starters are not deliberately added and lamb or kid rennet paste is used to coagulate the milk. These rennets are very rich in lipase and proteolytic enzymes. Milk is renneted at 32–38 °C for ca. 30–60 min. After renneting, the coagulum is broken into particles the size of rice grains. Broken

Fig. 6.119 Taleggio cheesemaking. Curd salting in brine. The photo has been kindly supplied by the Consorzio Taleggio PDO, Italy

curd rests under whey for 5 min. After removal from the vat, the curd grains are placed in molds, pressed manually to increase of whey drainage. During pressing, the cheeses rest in humid, warm rooms to help the milk autochthonous thermophilic lactic acid bacteria to complete the acidification process. To promote curd acidification after pressing, the curd is maintained in large warm wooden boxes at 30–40 °C and a relative humidity of 100%. At the end of this operation, the pH of the curd does not decrease below 5.3. The residual galactose is catabolized during ripening. Treatment of the curd with hot water or hot *scotta* is necessary to make the rind smooth, thick, and resistant. Salting is in brine (23–25°Bé) for 36–72 h at a temperature lower than 15 °C. Over the years, however, the intensity of salting has decreased. Ripening time ranges between a minimum of 3 months (for table cheese) and 6 months for cheese to be grated. During the early phase of ripening, smoking of the curd may be carried out by exposure to smoke from the wood of Mediterranean

Fig. 6.120 Taleggio cheesemaking. Curd ripening. The photo has been kindly supplied by the Consorzio Taleggio PDO, Italy

Fig. 6.121 Canestrato di Moliterno cheese. The photo has been kindly supplied by ONAF (Italian Cheese Taster Organisation, www.onaf.it)

scrub trees. During ripening, the cheese is rubbed with a mixture of olive oil and sheep fat.

The cheese has a cylindrical or wheel shape with curved sides, 12–18 cm high, 18–20 cm in diameter, and weigh 1.5–4.0 kg. The color of the cheese paste ranges between white and straw-yellow with a compact, hard, and friable structure. The cheese paste has no holes. The flavor and taste of Fiore Sardo are pronounced, aromatic, moderately spicy, and the rind varies from deep gold to dark brown, with a sour smell.

Fig. 6.122 Geographical area for the manufacture of Canestrato di Moliterno cheese

6.3.5   *Fossa of Sogliano al Rubicone*

The typical PDO Fossa (pit) cheese originated and is made in Sogliano al Rubicone (Cesena/Forlì province in the Emilia–Romagna region). Other provinces in the same region (Ravenna and Rimini) also manufacture the cheese (Mipaaf 2017; Mucchetti and Neviani 2006; Avellini et al. 1999; Gobbetti et al. 1999; Massa et al. 1988) (Figs. 6.129, 6.130, 6.131, and 6.132). Another but not PDO variety with only some technological similarities is manufactured in the Marche region (Urbino, Ancona, Ascoli Piceno, and Macerata provinces). The typical feature of this cheese is ripening in flask-shaped pits, which are dug in the tufa ground of Forlì-Cesena province. Cheesemaking is typically from raw ewe's milk but in several cases, mixed ewe–bovine milk is used. Traditionally, the cheese is produced only during a defined period of the year, from August to November.

Fig. 6.123 Canestrato Pugliese cheese. The photo has been kindly supplied by ONAF (Italian Cheese Taster Organisation, www.onaf.it)

The cheese is manufactured from whole raw or pasteurized milk. Natural milk thermophilic starters, consisting of selected milk autochthonous lactic acid bacteria, are used to inoculate the milk. The milk is coagulated using powdered calf rennet at 30–38 °C for 7–20 min. Curd breaking is into grains having dimensions of ca. 0.5–1.0 cm. The curd after molding is pressed and held at ca. 28 °C for 4–8 h. The cheese is salted using brine (18–22°Bé) or dry NaCl. Generally, the curd is ready for ripening in pits after a period (up to ca. 3 months, minimum 60 and maximum 240 days) of ripening in rooms, which is necessary to achieve a certain degree of consistency and eliminate the risk of whey losses when pressed into the pits. Before placing and pressing into the pits, cloths are used to clean and wrap the individual cheeses. The sides of the flask-shaped pits are covered with straw, which is fixed by wooden rods, horizontally linked through wooden rings. The pits are open during August and when completely filled with cheese, they are hermetically sealed. The humidity inside the pits is close to saturation and the temperature ranges from 17 to 25 °C. Traditionally, the pits are opened on November 28th; at this time, the ripening is at least for 6 months, including maturation in rooms. The cheese microbiota is extremely variable, mostly coming from milk autochthonous strains, and markedly influenced by the type of cheese technology used for cheesemaking (animal species that produced the milk, with or without the use of starters, with or without

Fig. 6.124 Geographical area for the manufacture of Canestrato Pugliese cheese

the use heat recovery treatments for milk). In the ripened cheese the amount of fat in the cheese dry matter is over 32% (Gobbetti et al. 1999).

Due to the pressure inside the pits, the shape of the cheese varies from cylindrical to very irregular, a rind is not present and the weight ranges from 0.5 to 1.9 kg. The color of the outer part of ripened cheese varies from ivory white to yellow amber. The presence of small fractures on the surface is part of the cheese characteristics. The inner cheese paste is semi-hard, friable, amber or slightly straw-colored. Most of the PDO cheeses do not have a clear distinction between paste and rind, so they are usually consumed without removing the outer part of the crust. The cheese flavor and taste vary according to the type of milk used. Ewe's cheese usually has an aromatic taste and a fragrant, intense and pleasing flavor, more or less accentuated. Cheeses made with a blend of ewe's milk and cow's milk are more delicate, moderately salty, and slightly acidic with a little bitterness.

Fig. 6.125 Casciotta di Urbino cheese. The photo has been kindly supplied by AFIDOP, Italy

6.3.6   Murazzano

Murazzano is a PDO cheese manufactured only in small territories around Cuneo (Piedmont region), particularly Murazzano and some little villages of the Comunità Montana Alta Langa di Bossolasco (Mipaaf 2017; Mucchetti and Neviani 2006) (Figs. 6.133 and 6.134). Murazzano is a soft-fresh cheese produced from ewe's milk alone or with a maximum addition of 40% cow's milk.

Raw, whole ewe's milk from one or two daily milkings is generally used, but thermized or pasteurized milk may be processed also. Traditionally, raw ewe's milk is not inoculated with a starter. Today, cooling milk to 4 °C is permitted and, consequently, pasteurization and natural milk or whey cultures or commercial/selected starters may be used. The milk is renneted at 37 °C for ca. 30 min by liquid calf rennet. After a first curd breaking into large pieces, the size of an orange, and after resting for ca. 10 min, part of the whey is removed and a second curd breaking is made to obtain grains the size of a hazelnut. The curd is molded and dry-salted after 24 h. Ripening lasts 4 to 10 days, and, sometimes, the curd is placed on cotton cloths. Ripening could be, exceptionally, extended to 2 months. During ripening, the curd is washed daily with warm water.

The shape of the cheese is cylindrical, 3–4 cm high and 10–15 cm in diameter, with a weight of 0.3–0.4 kg. Cheese paste structure is soft, finely granular, sometimes with a few holes and its color is white. The fresh cheese has no rind, with a slight light straw-colored surface for the little longer-ripened cheese. The mature cheese has a pleasant flavor, delicate and sweet for fresh cheeses, intense and sometimes with slight spicy notes for ripened cheeses.

Fig. 6.126 Geographical area for the manufacture of Casciotta di Urbino cheese

6.3.7   Pecorino Crotonese

Pecorino Crotonese is a PDO hard, low-cooked ewe's milk cheese manufactured in different territories around the cities of Crotone, Catanzaro, and Cosenza (Calabria region) (Mipaaf, 2017) (Figs. 6.135 and 6.136). Whole ewe's milk is used. Milk may be raw, thermized, or pasteurized. Natural milk or whey cultures or commercial/selected starters may be used. Renneting is done with kid rennet paste at 36–38 °C for ca. 40–50 min. The coagulum is broken into particles the size of rice grains. Maintaining curd grains under stirring, cooking is at 42–44 °C for 5–6 min. Then, the curd is left to compact on the bottom of the vat. After that, the bed of curd is cut into pieces and placed into the typical baskets to get the typical shape. The shaped curd undergoes further cooking by immersion for a few minutes in warm whey at a temperature less than 55 °C. This allows the optimal structure, whey

Fig. 6.127 Fiore Sardo cheese. The photo has been kindly supplied by AFIDOP, Italy

drainage, and rind formation. Salting is done using both dry salt and brine (a solution of water and sea salt). The duration of salting depends on the size of the cheese. The extent of ripening varies depending on the type of cheese. The cheeses are ripened in fresh and slightly ventilated rooms or, more traditionally, in sandstone caves. Usually, two types of cheese are produced, one ripened for 60 to 90 days, and another subjected to longer maturation than 90 days. The fat content in cheese dry matter should not be less than 40% and proteins should be higher than 25% (De Pasquale et al. in press).

The cheese is cylindrical in shape, 6–15 cm high and 10–20 cm in diameter, and weighs 0.5–5.0 kg. For the type ripened for more than 6 months, the cheese may weigh 10.0 kg. The cheese paste is white or light strawberry, with a uniform structure and the presence of a few holes. The rind is thin with typical signs from the basket used for molding. For the type subjected to long ripening, the rind becomes hard and brown, and may be rubbed with olive oil. A slight smell of ewe's milk interacts harmoniously with other odors of hay and herbs. The flavor is soft and slightly acidic, becoming more intense, with a slightly spicy aftertaste, with increasing ripening.

6.3.8 Piacentino Ennese

Piacentinu Ennese is a PDO pressed cheese manufactured from whole raw ewe's milk. It is a typical cheese from Sicily, where native sheep breeds Comisana, Pinzirita, and Valle del Belice are present. The area for cheesemaking is close to the city of Enna (Mipaaf 2017) (Figs. 6.137, 6.138, and 6.139).

Fig. 6.128 Geographical area for the manufacture of Fiore Sardo

Milk from one or two consecutive milkings is refrigerated and used for cheese-making. After warming to a maximum temperature of 38 °C, the milk is placed into a wooden container *tini*, where coagulation takes place. Before renneting, saffron is added to the milk (maximum level of 5 g per 100 L of milk). The milk is renneted with lamb or kid rennet paste at ca. 37–38 °C in ca. 45 min. The optimal texture is determined manually and by observing the whey drainage and color. Then, the gel is broken. To facilitate the drainage of curd grains, hot water at 75 °C is added (20 L per 100 L of milk). Breaking the curd proceeds until the particles have the size of rice grains. After extraction, the curd is placed on a wooden table, and roughly cut into pieces. Then, the curd is molded in the typical *canestro*. At this stage, grains of black pepper are added and pressing occurs to maximize the curd paste cohesion. The curd is removed and placed on wooden *tini* and covered with warm *scotta* for a period of ca. 3–4 h. Further, the cheese is allowed to dry at room temperature for

Fig. 6.129 Fossa of Sogliano al Rubicone.cheese The photo has been kindly supplied by ONAF (Italian Cheese Taster Organisation, www.onaf.it)

24 h. Salting is with dry salt, with repeated treatments for ca. 10 days. Ripening lasts for at least 60 days at 8–10 ° C and a relative humidity of 70/80%. The ripened cheese has a fat content in cheese dry matter not less than 40% and proteins not less than 35%. The maximum content of NaCl is 5% and the moisture content is not less than 30%.

The cheese is cylindrical in shape, 14–15 cm high and 20–21 cm in diameter, and weighs 4.5–4.5 kg. Cheeses ripened more than 6 months weigh ca. 10.0 kg. The cheese paste is homogeneously yellowish, smooth non-granular with a few small holes. The cheese has yellowish rind, more or less intense due to the presence of saffron. The rind shows the typical signs of the baskets used for molding and may be treated with olive oil. The thickness of the rind does not to exceed 5 mm. The cheese taste is delicate with a slight aroma of saffron. The flavor becomes more intense with ripening, and a slightly spicy aftertaste may be present.

Fig. 6.130 Geographical area for the manufacture of Fossa of Sogliano al Rubicone cheese

6.3.9   Pecorino Romano

Pecorino Romano is PDO extra-hard ewe's milk cheese manufactured in the Lazio region, within the territories around Rome, in Sardinia and around the city of Grosseto (Tuscany region) (Mipaaf 2017; AFIDOP 2016; Addis et al. 2015; Mangia et al. 2011; Manzi et al. 2007; Cecchi et al. 1998; Deiana et al. 1997; Deiana et al. 1984; Ledda and Arrizza 1969; Marcialis et al. 1968; Contardi 1965; Tomarelli 1951; Fascetti and Savini 1929; Besana 1915) (Figs. 6.140, 6.141, 6.142, 6.143, 6.144, 6.145, and 6.146).

Before cheesemaking, the raw ewe's milk may be stored for 48 h (24 h at the farm and 24 h at the dairy) at a temperature of 4–6 °C. Usually, the cheese is manufactured from raw or thermized milk, which is inoculated with a natural culture, *scotta fermento*, which is produced by acidifying the *scotta*, the deproteinized whey

Fig. 6.131 Pit for ripening of Fossa of Sogliano al Rubicone cheese. The photo has been kindly supplied by ONAF (Italian Cheese Taster Organisation, www.onaf.it)

obtained from the manufacture of Ricotta. Thermophilic lactic acid bacteria, such as *Streptococcus thermophilus*, *Lactobacillus delbrueckii* subsp. *lactis*, and *Lactobacillus helveticus*, dominate the microbiota of this natural starter. It is also possible to use some natural milk cultures or some commercial/selected starters isolated from milk originating from the regions of cheese production. The use of natural or selected starter became necessary after the introduction of themization of milk, useful to reduce the contaminant microbiota present in raw ewe's milk. The milk is coagulated using lamb rennet paste at 38–40 °C for ca. 30 min. The rennet paste is very rich in lipase and proteolytic enzymes. After cutting, the curds are cooked at 45–48 °C for ca. 10 min (Figs. 6.143 and 6.144). After removal from the vat, the curds are molded, pressed manually, and pierced by fingers or a stick or mechanically to increase whey drainage. During pressing, the cheeses are held in humid and warm rooms to help the thermophilic lactic acid bacteria to complete the

Fig. 6.132 Pit for ripening of Fossa of Sogliano al Rubicone cheese. The photo has been kindly supplied by Prof.ssa Rosalba Lanciotti (Bologna University)

Fig. 6.133 Murazzano cheese. The photo has been kindly supplied by ONAF (Italian Cheese Taster Organisation, www.onaf.it

acidification process. At the end of this fermentation, the pH of the cheese does not decrease below 5.3. The residual galactose in the curd is fermented during ripening. Salting is very important for this cheese variety, both for the laborious techniques used in the tradition and for the long length of the salting period, which may last up to 80–120 days.

Dry salting requires that the cheese is moistened with brine and then rolled over dry salt. The operation should be repeated after 5–10 days, salting the other face and continued with less frequency. When the cheeses are hard enough and able to withstand the pressure, they are coupled, or superimposed in piles or towers made up of

Fig. 6.134 Geographical area for the manufacture of Murazzano cheese

2, 3, or more cheeses and only face to face. The salting and drying time range between 60 and 100 days depending on cheese size. In the old tradition, cheeses could be pierced with a thin needle at 4–6 points in order to favor further whey drainage and a better penetration of the salt. To accelerate salting, a mixed technique may be used that involves immersion for 6 to 10 days in brine (23–24°Bé) at 11–13 °C, followed by dry salting based on 3–4 treatments during a period of 50–70 days. At the end of salting, the salt concentration in the cheese is particularly high and usually ranges from 4 to over 6%. The high concentration of NaCl in cheese favors the selection of a secondary microbiota in cheese during ripening, consisting of micrococci and non-pathogenic staphylococci and yeasts. Lactic starter bacteria from starter or raw milk are still present during ripening and drive the principal maturation biochemical processes. The cheeses are ripened for 8–12 months at 10–14 °C and a relative humidity of 70–80% to develop the charac-

Fig. 6.135 Pecorino Crotonese cheese. The photo has been kindly supplied by ONAF (Italian Cheese Taster Organisation, www.onaf.it)

teristic flavor (Figs. 6.145 and 6.146). The minimum ripening time is now set at 5 months for table cheese and 8 months for grating cheese. Ripening is characterized by lipolysis due primarily to the activity of the lipase (pregastric esterase) in lamb rennet paste. A high concentration of butyric acid is present, representing over 30% of the free fatty acids in ripened cheese. The presence of hexanoic and octanoic acids is also important for flavor intensity. The degree of proteolysis is not negligible, although the soluble nitrogen is always less than 30% of total nitrogen. The pH of the cheese rises during maturation and can reach 5.4–5.5.

The cheese has a cylindrical shape, is 25–40 cm high and 25–35 cm in diameter, and weighs 20–35 kg, depending on production technology. The cheese paste has a compact, hard, and friable structure and may contain little holes. The color is white or slightly straw-yellow. The rind is ivory white or light straw-colored and may be colored black when a protective film is used or if traditionally the cheese is treated with grape seed oil. The cheese has a distinctive taste with an aromatic slightly spicy flavor. Long-ripened Pecorino Romano cheeses are more or less spicy with an intense flavor.

6.3.10   Pecorino Sardo

This is a variety of Pecorino cheese, the manufacture of which is limited to several provinces of Sardinia. It is a table cheese and/or grated cheese produced exclusively from whole ewe's milk, low-cooked paste, with medium short maturation (Sweet Pecorino Sardo) or extended over two months (Ripened Pecorino Sardo) (Mipaaf

Fig. 6.136 Geographical area for the manufacture of Pecorino Crotonese cheese

2017; AFIDOP-IGP 2016; Mangia et al. 2013; Manzi et al. 2007; Mannu and Paba, 2002) (Figs. 6.147 and 6.148).

The milk may be stored at 4 °C before cheesemaking. Raw or thermized/pasteurized whole ewe's milk, natural *scotta fermento* or natural whey or milk cultures may be used. Differently from that observed for Pecorino Romano, coagulation is achieved using liquid or powdered calf rennet at 36–39 °C, and the overall coagulation time varies between 25 and 45 min. The coagulum is broken into grains with dimensions of ca. 0.5 cm, close to a hazelnut size and then cooked to 39–43 °C. A lower temperature of cooking is usually used for the sweet-type cheese. The curd is then removed from the vat and, when destined for the ripened type, is pressed in order to accelerate whey drainage. The curd is maintained in a humid warm environment (temperature, 35–45 °C and high relative humidity) in order to favor acidification and drainage of whey. Salting is carried out using both dry salt and saturated

Fig. 6.137 Piacentinu Ennese cheese. The photo has been kindly supplied by Prof. Germano Mucchetti (Parma University, Italy)

Fig. 6.138 Piacentinu Ennese cheese. The photo has been kindly supplied by ONAF (Italian Cheese Taster Organisation, www.onaf.it)

brine for ca. 48 h at 10–12 °C. Usually, the NaCl content does not exceed 2% in ripened cheese. The sweet type is ripened for 20 days, to a maximum of 60 days at 6–12 °C and a relative humidity of 80–85%. The hard variant is ripened for 2–8 months at 12 °C and a relative humidity of 80–85%. The microbiota of the cheese is predominantly represented by the natural cultures in the case of Pecorino Sardo obtained from thermally treated milk. In the case of Pecorino Sardo produced

Fig. 6.139 Geographical area for the manufacture of Piacentinu Ennese cheese

traditionally from raw milk the contribution of various milk autochthonous microbial species is observed.

The shape of the cheeses is cylindrical. The sweet type is 8–10 cm high and
15–18 cm in diameter, with a weight of 1.0–2.3 kg. The ripened type is 10–13 cm
high and 15–22 cm in diameter, with a weight of 1.7–4.0 kg. The color of the cheese
paste ranges between white and straw yellow with a compact elastic structure, for
the sweet type, hard and friable for the ripened type with little holes that may be
present. The straw yellow rind is initially smooth and springy, but later becomes
darker and harder. The taste is sweet for the youngest cheeses but it becomes a little
spicy for the longer-ripened ones. The mature ripened type has a pleasant pungent
flavor and a firm, hard, and granular texture.

Fig. 6.140 Pecorino Romano cheese. The photo has been kindly supplied by AFIDOP, Italy

6.3.11 *Pecorino Siciliano*

Pecorino Siciliano cheese is manufactured only in Sicily, between October and June, when whole ewe's milk of the zone is available. The PDO type is a hard cheese obtained using raw ewe's milk and ripened for at least 4 months (Mipaaf 2017; Mucchetti and Neviani 2006; Migliorisi et al. 1997; Candido 1994; Pirrone 1994; Patti et al. 1984, Patti et al. 1985; La Curlo 1925) (Figs. 6.149, 6.150, 6.151, and 6.152).

In the traditional method of production, cheese is produced from raw milk only and without addition of starters. Often, manufacture is directly on dairy farms and therefore, in that case, there is no collection and milk is not refrigerated. In the case of milk from two milkings, cooling of the evening milk occurs. In such technological conditions, considering the rich microbiota generally associated with ewe's milk, the microbial diversity of the cheese corresponds to that typical of milk, as modified by the particular cheesemaking parameters used and by the duration of ripening. Today, the manufacture may be also from thermized or even pasteurized milk. In these cases, natural milk or whey cultures, containing mainly thermophilic lactic acid bacteria, are the inoculum for the vat milk. Traditionally, a wooden vat, called *tina* is used. Renneting is done with lamb rennet paste which occurs within 40–60 min at a wide range of temperature (30–40 °C). The coagulum is broken up using a traditional wooden tool, known as a *rotella*, into pieces the dimensions of a pea. The curds are cooked at 40 °C for ca. 10 min by adding hot water (ca. 70–80 °C). Molding is in a circular vessel, traditionally called *canestro*, where the curd undergoes slight pressing. Salting is a point of difference in the production of different cheese types. In the case of cheese called *Azzima*, the curd may not be salted. In other cases, the salting is done by using both dry salt and brine (18–20Bé). The

Fig. 6.141 Geographical area for the manufacture of Pecorino Romano cheese

intensity of salting coupled to the ripening time defines the large range of products called Pecorino Siciliano: Primosale (one salty stage and ca. 20 days of ripening), semi-salted Pecorino (with two stages of salting and 2–3 months of ripening) and Pecorino Siciliano DOP with two stages of salting and at least 4 months of ripening. At the end of ripening, the fat content in the cheese dry matter of this last type must be more than 40%.

The DOP cheese has a cylindrical shape, 10–18 cm high and 35 cm in diameter, and weighs 4–12 kg. The cheese paste is compact, white or pale yellow, with little holes. The rind is yellowish-white and is much wrinkled due the basket molds used. The ripened cheese develops a moderate piquant flavor, which is distinctive and the aroma is intense. Pepato (peppery) is a variant of Pecorino Siciliano cheese, which differs by the addition of black pepper to the curd during molding.

Fig. 6.142 Pecorino Romano cheese. The photo has been kindly supplied by Consorzio Pecorino Romano PDO

Fig. 6.143 Pecorino Romano cheesemaking. Curd breaking. The photo has been kindly supplied by Consorzio Pecorino Romano PDO

6.3.12   Pecorino Toscano

The manufacture of PDO Pecorino Toscano cheese is limited to several provinces of Tuscany and some little territories in the Umbria and Lazio regions (Mipaaf 2017; AFIDOP-IGP 2016; Manzi et al. 2007; Mucchetti and Neviani 2006; Mannu and Paba 2002; Bizzarro et al. 2000; Bizzarro et al. 1999; Neviani et al. 1998; Tomarelli 1951) (Figs. 6.153, 6.154, 6.155, 6.156, 6.157, 6.158, 6.159, 6.160, 6.161, 6.162,

Fig. 6.144 Pecorino Romano cheesemaking. Curd grains after breaking. The photo has been kindly supplied by Consorzio Pecorino Romano PDO

Fig. 6.145 Pecorino Romano cheesemaking. Curd ripening. The photo has been kindly supplied by Consorzio Pecorino Romano PDO

6.163, 6.164, 6.165, 6.166, 6.167, 6.168, 6.169, and 6.170). It is a sweet or semi-sweet cheese, produced exclusively from whole ewe's milk from the geographical area of production, low cooked and short-ripened (ca. 20 days, Sweet Soft Pecorino Toscano) or medium-ripened (4 months, Ripened Pecorino Toscano). It is used as a table or grated cheese.

The whole ewe's milk may be stored before cheesemaking at a temperature of 4 °C, then it may be thermized or usually pasteurized. Natural milk cultures or starters selected from strains isolated and identified in milk or cheese manufactured in

Fig. 6.146 Pecorino Romano cheesemaking. Curd ripening. The photo has been kindly supplied by Consorzio Pecorino Romano PDO

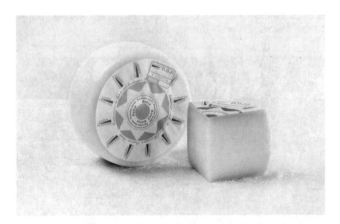

Fig. 6.147 Pecorino Sardo cheese. The photo has been kindly supplied by AFIDOP, Italy

the typical territory of production may be added to milk in the vat. Selected starters are used under the Consortium Pecorino Toscano control. Usually, starters consist of a mixture of thermophilic and mesophilic lactic acid bacteria with a role in cheese acidification and flavoring. The milk is renneted using liquid or powdered calf rennet, or rarely vegetable rennet, at 33–38 °C and the overall coagulation time varies between 20 and 30 min. Breaking the coagulum is one of the basic parameters that differentiate between the sweet soft and the ripened cheeses (Figs. 6.155, 6.156, 6.157, and 6.158). For the sweet soft cheese, the coagulum is broken into grains the size of hazelnuts or maize, while for the ripened cheese it is finely broken to drain more whey quickly. This diversity in the intensity of breaking mode may also explain the different quantities of fat that remains in the different types of cheese, which ranges from 45% for sweet soft cheese to 40% for the ripened cheese. For

Fig. 6.148 Geographical area for the manufacture of Pecorino Sardo cheese

Ripened Pecorino Toscano, previously the broken curd was cooked to 40–42 °C for 10–15 min, but nowadays this practice is rarely used. A lower cooking temperature may be used to give a softer and more elastic cheese paste. Depending on the size and the mechanization of the dairy, removal of the curd from the vat for molding could be manual or by mechanical pumping and mechanization of the processes (Figs. 6.159, 6.160, 6.161, 6.162, 6.163, 6.164, 6.165, and 6.166). Molded curd may be partially pressed and maintained in a warm humid environment. The curd is stored in a warm room or in large steel or wooden boxes heated by steam injection to reach, even, 50 °C for a maximum of 10 h, in order to obtain different intensity in cheese acidification and drainage of whey. This operation is also useful for preventing the early formation of a thick rind. The whole process may be performed using an air-conditioned room at about 40 °C, instead of the big steel boxes. If a mesophilic starter has been used, a lower temperature of heating may be used, but usually

Fig. 6.149 Pecorino Siciliano cheese. The photo has been kindly supplied by Prof. Germano Mucchetti (Parma University, Italy)

never lower than 32 °C. At the end of heated storage, the pH of the cheese paste is usually close to 5.0. Acidification of the curd to a pH of ca. 5.0 in the case of cheese produced with thermophilic starters can be completed within about 3 h after removing the curd from the vat. The characteristic microbiota is represented by the starter thermophilic lactic acid bacteria, composed mainly of various *Streptococcus thermophilus* strains. *Lactococcus* biotypes used as starters, and non-starter lactobacilli that survive the heat treatment of milk, represent largely the mesophilic biota. Salting is done using both dry salt and brine (18–20°Bè) (Figs. 6.167 and 6.168). Usually, the salt content of the ripened cheese does not exceed 1.5–1.7%.

The cheese has a cylindrical shape, 7–11 cm high and 15–22 cm in diameter, and weighs 0.75–3.5 kg (Figs. 6.169 and 6.170). The cheese paste is hard in ripened cheese and semi-hard for the sweet soft type. The straw-yellow rind during ripening becomes darker and harder. Rarely, the rind undergoes treatment with oil, ash, tomato juice or peelings, or using surface coatings that reproduce colors derived from traditional treatments. The mature cheese has a pleasant sweet flavor that for the ripened type could become more pungent, aromatic, and even moderately spicy as is usual for many ewe's milk cheeses.

6.4   Pasta Filata Cheeses

The stretched-curd cheeses, in Italian *pasta filata* cheeses, represent a particular variety of Italian cheeses characterized by the heating and stretching of acidified curd (Fox et al. 2017; Mucchetti et al. 2017; Mucchetti and Neviani 2006).This

Fig. 6.150 Geographical area for the manufacture of Pecorino Siciliano cheese

technological process markedly modifies the structure of the casein and sensory perception of the curd.

Pasta filata cheeses may be produced from cow's or water buffalo's milk. Contrarily to cheeses more or less long-ripened (e.g., Scamorza, Provolone), the fresh *pasta filata* cheeses are usually stored in protective liquid (salted whey/water also with added organic acids) (e.g., Mozzarella, Burrata).

Raw or pasteurized milk may be used, depending of the cheese type to be produced. This choice involves the residual microbiota present in the curd and in the ripened cheese. The only common feature of *pasta filata* cheeses is the stretching obtained by heating the acidified curd. This technological operation modifies the structure of the properly demineralized acid curd by the addition of hot water (85–95 °C or less), which induces the first phase of curd melting. The final stretching of the melted curd is done manually or mechanically. Demineralization of the curd

Fig. 6.151 Pecorino Siciliano cheese. The photo has been kindly supplied by ONAF (Italian Cheese Taster Organisation, www.onaf.it)

Fig. 6.152 Pecorino Siciliano cheese. The photo has been kindly supplied by ONAF (Italian Cheese Taster Organisation, www.onaf.it

may be the result of the acidifying activity of starter lactic bacteria (as done for traditional production) or from sequestering colloidal calcium bound to caseins through acidification of the milk before coagulation by addition of organic acids (usually citric or lactic acid). These two processes may even be used combined. Preacidification by adding organic acids, coupled with milk pasteurization, usually it is not used to produce long-ripened *pasta filata* cheeses. The milk fermentation by

Fig. 6.153 Pecorino Toscano cheese. The photo has been kindly supplied by Cosorzio Pecorino Toscano

lactic acid bacteria starters reduces the pH of curd. The pH necessary for stretching, ranges from 5.2 to 4.8, depending on the type of curd obtained and on the milk used. The time needed to achieve this acidification depends on the metabolic activity of the starters. Consequently, this time may vary within a wide range, also depending on the different technological options used to produce the curd (e.g., pH, temperature, size of the curd). Preacidification of milk by adding citric acid solutions allows solubilization of a large part of the total milk calcium (about 40–60%). In this case, the pH of the milk, which results in a proper demineralization of the curd, ranges from 5.5 to 5.9. These values of pH correspond approximately the same level of demineralization of caseins obtained at a lower pH with biological acidification. The essential phenomenon that allows curd to be stretched on heating is the loss of colloidal calcium in the casein. The stretching of demineralized curd is then the result of (1) cutting the curd, (2) curd melting, and (3) stirring the molten curd. In order to be stretched, the demineralized curd is first melted by addition of hot water (85–95 °C or less). The intensity of thermal treatment (time/temperature) also determines (1) the degree of inactivation of the proteolytic activity of rennet and (2) the reduction of the contaminating microbiota and lactic acid bacteria starters, which may result also as a consequence of the pH reduction. The melted and stirred curd assumes the desired shape and size. Traditionally, shaping was done by hand, but today it is often performed with molding machines. A smoking step may be included also. In the case of fresh *pasta filata* cheeses, the choice of packaging and the immersion in protective liquid (salted whey/water also containing added organic acids) are essential in defining the outcome of the process (De and Gobbetti 2011).

Fig. 6.154 Geographical area for the manufacture of Pecorino Toscano cheese

6.4.1 *Mozzarella*

Mozzarella cheese has worldwide distribution, probably as a consequence of the migration of Italian people in the last centuries. Nowadays, the name Mozzarella includes several types of Italian fresh pasta filata cheeses (Mucchetti et al. 2017; Mucchetti and Neviani 2006; Addeo et al. 1997a, b).

According to Codex Alimentarius, Mozzarella may be defined as an unripened, rindless, smooth, and elastic cheese with a long-stranded parallel-oriented fibrous protein structure without evidence of curd granules. In conclusion, Mozzarella is a soft cheese with overlying layers.

It possible to use curd produced previously (in the same dairy or elsewhere). Usually, the milk to be processed is stored after milking at 4–6 °C. The fat content of the milk may be adjusted to obtain its desired content in the cheese. The milk may

Fig. 6.155 Pecorino Toscano cheesemaking. Milk placed in the vat. The photo has been kindly supplied by the Consorzio Pecorino Toscano PDO

Fig. 6.156 Pecorino Toscano cheesemaking. Curd breaking. The photo has been kindly supplied by the Consorzio Pecorino Toscano PDO

Fig. 6.157 Pecorino Toscano cheesemaking. Curd breaking. The photo has been kindly supplied by the Consorzio Pecorino Toscano PDO

Fig. 6.158 Pecorino Toscano cheesemaking. Curd breaking. The photo has been kindly supplied by the Consorzio Pecorino Toscano PDO

Fig. 6.159 Pecorino Toscano cheesemaking. Curd after extraction, and before drainage and molding. The photo has been kindly supplied by the Consorzio Pecorino Toscano PDO

be raw or heat-treated (in this case usually pasteurized). Different types of starter may be added to the vat milk: natural whey cultures, natural milk cultures, or commercial/selected starters. The type and amount of rennet used, breaking the coagulum, and recovery of the curds may be very different. After acidification (in cases when a starter is used), the curd is heated and stretched. In some cases, NaCl is added to the stretching water. The stretching of the acid curd is an operation, which induces a severe heat stress on microorganisms present, reducing their number and their viability according to their heat tolerance. After stretching the melted curd, it is shaped by hand or mechanically still at an elevated temperature (85–95 °C or less). The cheese is hardened by cooling, usually by dipping into fresh water, and then eventually placed in brine. Cooling with cold water and brining are two steps that could cause potential post-contamination of cheese, particularly on the surface. The gross chemical composition of different varieties of Mozzarella cheese shows

Fig. 6.160 Pecorino Toscano cheesemaking. Curd molding and drainage. The photo has been kindly supplied by the Consorzio Pecorino Toscano PDO

that the main component is water, almost always over 56%, except for Mozzarella cheese used for making pizza.

The cheese may be formed into various shapes and sizes and has a near white color. It is characterized by a milky fresh taste and a high moisture content, with little or no proteolysis. The structure of the cheese paste also largely depends on the method used for curd stretching (manual or mechanical). It is usually packed and immersed in a solution of protective liquid although the largest sizes may be packed without liquid.

The several types of Mozzarella cheese may be classified as follows: (1) Mozzarella di Bufala Campana PDO, from water buffalo milk produced in a specified area of Southern Italy, with the use of a natural whey culture; (2) Mozzarella

Fig. 6.161 Pecorino Toscano cheesemaking. Curd molding and drainage. The photo has been kindly supplied by the Consorzio Pecorino Toscano PDO

Fig. 6.162 Pecorino Toscano cheesemaking. Curd molding. The photo has been kindly supplied by the Consorzio Pecorino Toscano PDO

from buffalo's milk, not PDO, often produced with a commercial/selected starter in a larger production geographical area; (3) traditional Mozzarella from cow's milk, obtained using a commercial/selected lactic acid bacteria starter, composed mainly of *Streptococcus thermophilus*, with the presence of enterococci and other non-spore-forming unidentified, heat-resistant bacteria; (4) Mozzarella cheese in light brine, obtained by natural milk or whey cultures or commercial/selected lactic acid bacteria starters or by direct acidification; (5) Pizza cheese, used as an ingredient for pizza; and (6) Mozzarella cheese obtained from a blend of cow and buffalo milk.

Fig. 6.163 Pecorino Toscano cheesemaking. Curd molding. The photo has been kindly supplied by the Consorzio Pecorino Toscano PDO

Fig. 6.164 Pecorino Toscano cheesemaking. Curd molding. The photo has been kindly supplied by the Consorzio Pecorino Toscano PDO

6.4.2 *Mozzarella di Bufala Campana*

The manufacture of this PDO cheese is limited to several provinces in the South of Italy (Caserta, Salerno, Napoli, Benevento, Frosinone, Latina, Roma, Foggia, and Isernia) (Mipaaf 2017; Mucchetti et al. 2017; AFIDOP-IGP 2016; Romano et al. 2011; Manzi and Pizzoferrato 2009; Manzi et al. 2007; Mucchetti and Neviani 2006; Mauriello et al. 2003; Ercolini et al. 2001; Mauriello et al. 2001; Coppola et al. 1988; Addeo et al. 1997a, b; Mucchetti et al. 1997; Moio et al. 1993) (Figs. 6.171, 6.172, 6.173, 6.174, 6.175, 6.176, and 6.177). Mozzarella di Bufala Campana (water buffalo milk Mozzarella) is a fresh pasta filata cheese, produced

Fig. 6.165   Pecorino Toscano cheesemaking. Curd held in warm room. The photo has been kindly supplied by the Consorzio Pecorino Toscano PDO

exclusively from buffalo milk and preserved by immersion in protective liquid (keeping liquid, including water and sometimes soft brine).

Whole buffalo milk, with a minimum fat content of 7.2%, is used. The milk used may be raw or pasteurized and is inoculated with a thermophilic natural whey culture starter. Different biotypes of *Streptococcus thermophilus*, *Lactococcus lactis*, *Lactobacillus delbrueckii spp. lactis*, and *Lactobacillus helveticus* species are usually present. The milk is renneted using liquid rennet at a temperature between 33 and 39 °C. After ca. 60–80 min, the coagulum is broken and the curd held under whey for a time close to 5 h. During this phase, a part of the whey may be removed and partially replaced by warm whey. This practice avoids the excessive cooling of the whey and favors the activity of the microbiota of the natural whey culture starter. The curd is then separated from the whey (Fig. 6.175). When the pH of curd reaches 4.7–4.9, it is ready for stretching and the curd is cut in big strips, and placed on a table to finish drainage. The curd is then cut into little pieces and mixed with water at ca. 95 °C. In this way, the curd initially reaches a temperature exceeding 75 °C and usually remains at a temperature above 68 °C for over 3 min. The length of the manual stretching phase is ca. 5 min (Fig. 6.173). During stretching, besides modification of the structure, there are changes in chemical composition. The curd looses soluble and fatty substances to the stretching water and absorbs warm water. The curd is differently shaped, by hand or mechanically (Figs. 6.174 and 6.176), and cooled to reach the firm desired texture and salted in brine. The local product is generally consumed within 24 h of production, while that used for more distant markets may be stored for up to 15–20 days. The composition of the cheese microbiota is determined by the production technology used (use of raw or heat-treated milk and methods of obtaining natural starter) with particular attention to stretching temperature that may give similar effects to pasteurization. A maximum of 65% moisture is acceptable.

Fig. 6.166. Pecorino Toscano cheesemaking. Curd held in warm room. The photo has been kindly supplied by the Consorzio Pecorino Toscano PDO

The most common form of this cheese is the rounded form, but also the braided or nodular form is commonly present in the market (Fig. 6.177). The size ranges between 10 and 800 g, but usually between 100 and 250 g. The texture of the cheese paste is soft and characteristic of pasta filata cheeses, with overlying layers that may form pockets containing liquid of milky appearance. When cut, an abundant release of whey must occur. This skin-rind has a smooth, nonviscous or flattened surface, with a thickness less than 1 mm and has a porcelain white color. The taste and flavor, reminiscent of milk, are characteristic and delicate.

Fig. 6.167 Pecorino Toscano cheesemaking. Curd dry salting. The photo has been kindly supplied by the Consorzio Pecorino Toscano PDO

6.4.3 Burrata di Andria

Burrata di Andria is a PGI *pasta filata* cheese produced from raw or pasteurized cow's milk. The soft exterior of the cheese (bag) consists exclusively of stretched paste, like mozzarella paste, while the inner part (core), called *stracciatella*, is a mixture of cream and Mozzarella cheese strips. The area of manufacture includes the entire territory of the Apulia region (Minervini et al. 2017; Mipaaf 2017; Mucchetti and Neviani 2006) (Figs. 6.178 and 6.179).

The gross chemical composition of the cow's milk used for cheesemaking is strictly defined. The cream used may be obtained by raw milk or whey centrifuga-

Fig. 6.168 Pecorino Toscano cheesemaking. Curd held in brine with special cages. The photo has been kindly supplied by the Consorzio Pecorino Toscano PDO

tion, in this case the separated cream must be pasteurized (72 °C for 15 s). The use of commercial pasteurized or UHT (Ultra High Temperature) cream is permitted also. The manufacture using pasteurized cow's milk, which is mixed with the acidified whey of the previous-day cheesemaking to give an initial pH of ca. 6.1, is another technology option. The manufacture also includes the optional use of natural milk or whey cultures or commercial/selected starters. The milk is renneted with calf rennet at 32–37 °C for 15–30 min. After obtaining the *pasta filata*, it is separated into two parts manually. The first part is shaped like a *sacco* (bag) to contain the second one. This latter is mixed with the cream to fill the bag. Closing the *sacco* is by warm water to get the desired shape. Salt could be added directly during

Fig. 6.169 Pecorino Toscano cheesemaking. Curd ripening. The photo has been kindly supplied by the Consorzio Pecorino Toscano PDO

Fig. 6.170 Pecorino Toscano cheesemaking. Ripened cheese. The photo has been kindly supplied by the Consorzio Pecorino Toscano PDO

stretching or the shaped curd could be placed in brine. The cheese has a moisture content between 60 and 70%.

The cheese looks like a spherical bag, which includes the core made of a mixture of cheese paste and cream. The term *stracciatella*, also used for this cheese, refers precisely to the mixture of the core content. The cheese weighs from 0.1 to 1 kg. The cheese paste is white and the outer stretched paste has a thickness of ca. 2 cm. The flavor and taste resemble that of milk and are more delicate on the surface of the product, increasing the buttery note as it approaches the center.

Fig. 6.171 Mozzarella di Bufala Campana cheese. The photo has been kindly supplied by AFIDOP, Italy

6.4.4 Vastedda del Belice

Vastedda del Belice is a DOP fresh *pasta filata* cheese made exclusively or mainly from ewe's milk. Vastedda is a small round cheese without rind, characterized by the unusual shape of a bun. The geographical area of sheep breeding, milk production, and processing is in the Belìce Valley (Agrigento, Trapani, and Palermo provinces) (Mipaaf 2017; Mucchetti et al. 2008; Mucchetti and Neviani 2006) (Figs. 6.180 and 6.181).

Knowledge about Vastedda cheese technology and composition is scarce. Usually, milk from one or two milkings must be processed within 48 h from the first milking. It is possible to refrigerate the milk. The milk is renneted with lamb rennet paste at 36–40 °C, in 40–50 min. The coagulum is broken into particles the size of rice grains. Syneresis is favored by adding hot water during curd breaking. Curd grains deposited at the bottom of the vat, are left to rest for 5 min to favor their cohesion and then the mass is deposited on rushes without pressing. Under these conditions, the curd stays at room temperature for a variable time (24–48 h). The length of this operation depends on the level of acidification, which must reach pH of ca. 4.7–5.5. The curd is then cut into slices, placed in a wooden container, called *piddiaturi*, and covered with hot water at a temperature of 80–90 °C. After stretching, the *pasta filata* is placed on ceramic plates, where, after being turned, it assumes the characteristic shape. After 6–12 h, salting is in saturated brine for 30 min to 2 h. Despite the low amount of cheese production and the low number of producers, there is no agreement about the methodology for salting. Cheese drying follows in fresh and moderately ventilated rooms. After 12–48 h, the cheese may be consumed.

Fig. 6.172 Geographical area for the manufacture of Mozzarella di Bufala Campana cheese

The fat content of the cheese must be higher than 35% in cheese dry matter. The content of NaCl must be less than 5% in cheese dry matter.

The cheese has a cylindrical shape, with slightly convex faces, 3–4 cm high and 15–17 cm in diameter; the weight ranges between 0.4 and 0.6 kg. The cheese paste is white without a crust. The taste is characteristic of fresh ewe's milk. The flavor is of fresh ewe's milk cheese with mild acidic notes. Traditionally, Vastedda is a fresh cheese, consumed within a few days of the cheesemaking mainly in the production area. Nowadays, thanks to the growing interest in traditional foods, Vastedda is marketed also outside Sicily. To allow its marketing, and to prolong its shelf-life, the cheese immediately after production is sealed under vacuum in water vapor-impermeable plastic film and stored at 4 °C for up to 60 days.

Fig. 6.173 Mozzarella di Bufala Campana cheesemaking. Curd stretching. The photo has been kindly supplied by Prof. Germano Mucchetti (Parma University, Italy)

Fig. 6.174 Mozzarella di Bufala Campana cheesemaking. Curd shaping. The photo has been kindly supplied by Prof. Germano Mucchetti (Parma University, Italy)

6.4.5 *Provolone and Other Ripened* Pasta Filata *Cheeses*

Cheeses with the name Provolone are generally different types of ripened *pasta filata* cheese, produced in different territories of Italy (Mucchetti and Neviani 2006; Barzaghi et al. 1997; Neviani et al. 1992; Addeo and Laezza 1987; Resmini et al. 1987; Zamorani et al. 1987; Battistotti et al. 1986; Carini et al. 1985; Battistotti

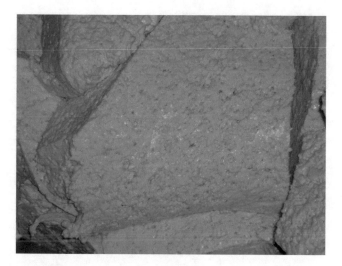

Fig. 6.175 Mozzarella di Bufala Campana cheesemaking. Curd before stretching. The photo has been kindly supplied by Prof. Germano Mucchetti (Parma University, Italy)

Fig. 6.176 Mozzarella di Bufala Campana cheesemaking. Curd mechanical shaping. The photo has been kindly supplied by Prof. Germano Mucchetti (Parma University, Italy)

et al. 1980; Bodini 1973; Bottazzi and Battistotti 1973; Ghitti 1959; Emaldi 1960). Originally, ripened pasta filata cheeses, like Provolone, were typical cheeses from southern Italy. Since the first half of the nineteenth century, in particular in the case of Provolone, they were produced also in the northern of Italy, particularly in the Po Valley. Different traditional names are used for these ripened pasta filata cheeses, referring to the different shapes and sizes in which they are prepared (e.g., Caciocavallo, Mandarino, Provolone, Pancetta, Pancettone, Scamorza).

Fig. 6.177 Mozzarella di Bufala Campana. The photo has been kindly supplied by Prof. Germano Mucchetti (Parma University, Italy)

Fig. 6.178 Burrata di Andria cheese. The photo has been kindly supplied by ONAF (Italian Cheese Taster Organisation, www.onaf.it)

The cheeses may be produced in two types, sweet and spicy. Usually, manufacture is with whole cow's milk, which may be raw, thermized, or pasteurized. Raw milk is mainly used to produce long-ripened spicy types. Vat milk is inoculated with a starter, usually a natural whey culture. The milk is renneted using liquid calf rennet or rennet paste for the spicy type. The coagulum is broken into different sizes and held to acidify. After reaching the necessary acidity, it is stretched and shaped. Then, the curd is salted in brine and in some cases smoked. Ripening may last for

Fig. 6.179 Geographical area for the manufacture of Burrata di Andria cheese

different times, depending on the cheese size; it may range from some days for little cheeses to more than one year for the large ripened cheeses. During ripening, various proteolytic and lipolytic events occur. Lipolysis may be really deep in the case of spicy types produced using rennet paste.

These cheeses are made in various shapes and may be spherical, pear shaped, cylindrical or even parallelepiped. Their weight ranges from less than 50 g to more than 10 kg and may reach 100 kg in certain cases. The characteristic shape of Provolone is truncated-conic, with shallow crevices for the passage of the ropes used to hang the cheeses coupled in pairs. The color of cheese paste is usually straw-yellow. The cheese paste is compact and elastic with the possible presence of a few small holes. Cheeses may have a thin crust, smooth, straw-yellow or close to brown if smoked. The flavor and taste depend mainly on the cheesemaking technology used and cheese size. Usually, flavor and taste range from delicate or little pronounced characteristic of the sweet type, not ripened, to intense and spicy.

Fig. 6.180 Vastedda del Belice cheese. The photo has been kindly supplied by Prof. Germano Mucchetti (Parma University, Italy)

6.4.5.1 Provolone Valpadana

Provolone Valpadana is a PDO *pasta filata* ripened cheese produced exclusively from cow's milk) (Mipaaf 2017; Mucchetti et al. 2017; AFIDOP-IGP 2016; Manzi et al. 2007; Mucchetti and Neviani 2006) (Figs. 6.182, 6.183, 6.184, 6.185, 6.186, 6.187, 6.188, 6.189, 6.190, 6.191, 6.192, and 6.193). The manufacture of this cheese is limited to several provinces close to the Po Valley (Brescia, Bergamo, Cremona, Verona, Vicenza, Rovigo, Padova, Piacenza, Mantova, Milano, Lodi, and Trento). It is principally distinguished as two types, Provolone Valpadana sweet and spicy. The cheese may be smoked.

Usually, whole cow's milk is stored at 4 °C after milking and sometimes the warm morning milk is mixed with the refrigerated evening milk. Raw milk is traditionally used for the spicy type, while pasteurized milk is used for sweet type. Bactofugation of milk may be used also and the addition of hexamethylenetetramine to the curd during stretching is used to inhibit undesired fermentation. A natural whey culture starter is used, in some cases enriched with some commercial/selected strains. The natural whey culture starter is predominantly made up of an association of *Lactobacillus helveticus*, *Lactobacillus delbrueckii* sp. *bulgaricus*, and *Streptococcus thermophilus* strains, which may be present in different numerical ratios. The natural whey culture is obtained by incubating whey from a previous cheesemaking, separated from the curd after the first or second cooking. Liquid calf rennet for the sweet type and lamb and/or kid rennet paste for the spicy type is used. The milk is renneted at 36–40 °C for 13–20 min (Figs. 6.184 and 6.185). Initially, the broken coagulum is heated to ca. 45 °C for 3–10 min using an amount of warm whey previously separated (whey temperature ranging between 65 and 85 °C, this separated and heated whey is called *calda*). Then, the curd is cooked using another

Fig. 6.181 Geographical area for the manufacture of Vastedda del Belice cheese

warming step, always using another *calda,* to reach 50–53 °C for ca. 3–6 min. For industrial production using large and automated vats the curd cooking could be done in only one warming step. The broken curd is placed on tables at a controlled temperature of 40–45 °C to complete whey drainage and curd acidification. The duration of the acidification phase of the curd varies on average from 4 to 6 h, exceptionally up to 18 h (Figs. 6.186 and 6.187). The curd is ready for stretching when it reaches pH 5.2–4.9. Under this condition, much of the calcium associated with casein is solubilized in the serum phase of the curd. In the case of the spicy type, the pH reached could decrease to 5.0–4.8. The acidified curd is cut and heated using warm water to reach 65–90 °C. The stretching phase is completed in 5–15 min (Fig. 6.188). The hot, and hence extremely plastic, paste is manually or mechanically formed according to the sizes and shape desired and then immediately cooled in water at 10–15 °C for 12–20 h to maintain the shape and consolidate the texture

Fig. 6.182 Provolone Valpadana cheese. The photo has been kindly supplied by AFIDOP, Italy

(Figs. 6.189 and 6.190). In the case of the cheese shaped like *mandarone*, the newly formed paste is tied to keep the typical shape. Salting is usually in brine (16–20°Bé) (Fig. 6.191) and then ripened for different times according to size and type of cheese (sweet or spicy) at 12–18 °C and a relative humidity of 85–90%. The ripening of sweet type is related mainly to proteolysis, by starter lactic acid bacteria and milk plasmin, while the ripening of the spicy type is strongly marked by lipolysis due to the use of rennet paste. The cheese microbiota is composed mainly of the lactic bacteria of the natural whey culture starter and of raw milk, as selected by curd cooking and stretching. The maximum water content must not exceed 46% for all sweet and spicy types up to 6 kg, and more than 43% for the spicy type above 6 kg. The fat in the dry matter must be in the range 44–54%.

Many different cheese shapes are possible (Figs. 6.192 and 6.193). The most popular are salami (or bacon), mandarin, and pear with spherical head or flask (classic Caciocavallo). Also the weigh may be very different. Ripening time depends on the size and shape of the cheese and may vary as follows: (1) up to 6 kg, minimum 10 days; (2) over 6 kg, minimum 30 days; (3) over 15 kg and only for spicy type, minimum 90 days; and (4) over 30 kg spicy type, aged for over 8 months. The texture of the cheese is related to its size and to the ripening time. The color of the cheese paste is almost white or slightly straw-colored in young cheese that can become strawberry after long ripening. Cheese paste is compact, although the presence of little holes is accepted and even the presence of a few openings (slots) that are more evident in ripened cheeses. The rind is smooth. The flavor ranges from delicate or little pronounced, characteristic of the sweet type, not ripened, to intense and spicy, due to the proteolysis and lipolysis caused by the use of rennet in lamb or kid paste. Provolone is often covered with wax or paraffin to avoid mold growth.

Fig. 6.183 Geographical area for the manufacture of Provolone Valpadana cheese

6.4.5.2   Provolone del Monaco

Provolone del Monaco is a PDO *pasta filata* ripened cheese, produced exclusively from cow's milk (Mipaaf 2017) (Figs. 6.194 and 6.195). The manufacture is limited to several areas in the province of Naples (e.g., Agerola, Casola di Napoli, Castellammare di Stabia, Gragnano).

Traditionally, raw cow's milk is used. After milking, refrigeration of the milk is allowed and manufacture has to start within 48 h from milking. Natural whey cultures may be used. Renneting is done with liquid calf rennet and/or kid rennet paste alone or in a mixture with at least 50% of the second one. Coagulation takes place at 32–42 °C in 40–60 min. The coagulum is broken into very small particles of hazelnut size and then of a maize grain size. Initially, curd grains undergo heating at ca. 48–52 °C, for a maximum of 30 min at ca. 45 °C. The broken curd is, then,

Fig. 6.184 Provolone Valpadana cheesemaking. Vat for cheesemaking. The photo has been kindly supplied by the Consorzio Provolone Valpadana PDO

Fig. 6.185 Provolone Valpadana cheesemaking. Curd grains after breaking. The photo has been kindly supplied by the Consorzio Provolone Valpadana PDO

placed on tables to complete whey drainage and acidification. The acidified curd is cut further and heated using water at 85–95°. When stretching is completed, the curd is cooled in water. Usually, salting is in brine (18–20°Bé) for a time depending on the size of the cheese (e.g., 8–12 h per kg). Ripening lasts at least 6 months. The fat content of the ripened cheese must exceed 40.5% of the cheese dry matter.

The cheese has a shape like a watermelon or pear. The weight varies from 2.5 to 8 kg. The cheese paste has a creamy color with yellowish tones, the texture is

Fig. 6.186 Provolone Valpadana cheesemaking. Curd extraction. The photo has been kindly supplied by the Consorzio Provolone Valpadana PDO

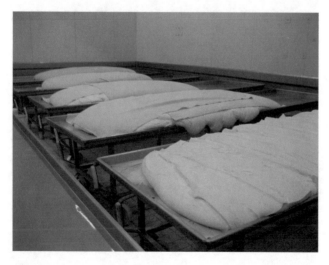

Fig. 6.187 Provolone Valpadana cheesemaking. Curd acidification. The photo has been kindly supplied by the Consorzio Provolone Valpadana PDO

elastic, compact, and uniform. Little holes, with a diameter of ca. 5–12 mm, are present, most abundantly towards the center of the cheese. The rind is thin and yellowish with slightly darker tones, almost smooth with light longitudinal grooves used to support cheeses during ripening. The flavor is delicate and little spicy.

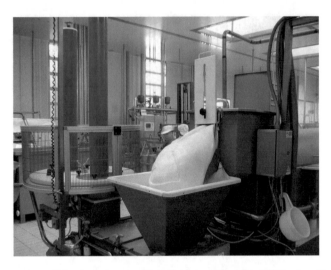

Fig. 6.188 Provolone Valpadana cheesemaking. Curd stretching after acidification. The photo has been kindly supplied by the Consorzio Provolone Valpadana PDO

Fig. 6.189 Provolone Valpadana cheesemaking. Curd shaping. The photo has been kindly supplied by the Consorzio Provolone Valpadana PDO

6.4.5.3   Ragusano

Ragusano is a PDO *pasta filata* ripened, semi-hard cheese with cooked paste, produced exclusively from raw whole cow's milk (Mipaaf 2017; Mucchetti and Neviani 2006; Fazzino et al. 2002; Randazzo et al. 1998, 2002; Licitra 2001; Licitra et al. 2000; Fulco et al. 1984; Carini and Baglieri 1981; Zamorani et al. 1974). Manufacture

Fig. 6.190 Provolone Valpadana cheesemaking. Curd shaping. The photo has been kindly supplied by the Consorzio Provolone Valpadana PDO

Fig. 6.191 Provolone Valpadana cheesemaking. Curd salting in brine. The photo has been kindly supplied by the Consorzio Provolone Valpadana PDO

is limited to some Sicilian territories in Ragusa and Siracusa provinces (Figs. 6.196 and 6.197).

Usually, whole cow's milk from a single milking is processed warm, but in some cases, it may be stored at 4 °C after milking. Traditionally, raw milk is used without addition of starters. Milk and curd fermentation, necessary to reach the acidity required for stretching, is by milk autochthonous lactic acid bacteria. Selection of the cheese microbiota is obtained only as a consequence of the technological param-

Fig. 6.192 Provolone Valpadana cheesemaking. Curd ripening. The photo has been kindly supplied by the Consorzio Provolone Valpadana PDO

eters chosen. Lamb and/or kid rennet paste diluted in brine is used. The milk is renneted at 34–37 °C for 60–80 min. Once, renneting was done in traditional little wooden tanks, called *tina*, but today different technological choices are possible. The coagulum is broken rapidly (a few minutes) and curd grains reach the dimensions of lentil. The curd is cooked using warm water added to reach a temperature close to 45 °C, which favors the growth of the raw milk thermophilic microbiota. The curd is held under warm diluted whey to favor acidification necessary for stretching. The broken curd may also be placed on tables at a controlled temperature to complete whey drainage and curd acidification. The curd is cut into thin slices and placed in a wooden or copper container (called the "staple") and covered with warm water at ca. 80 °C, with the ratio of about 1:1 with the curd, for 4–10 min. The water temperature, the contact time, and the amount of water vary depending on the acidity reached in the curd, so that less acidic and demineralized curd is provided with less heat and vice versa. The *pasta filata* is, then, formed manually in spherical form and pressed to reach the typical shape (using *mastredda*). Cooling is performed directly in *mastredda* without using cold water, and usually requires a long time. Cheeses are usually salted in brine possibly coupled with the use of dry salt. Ripening time ranges from 5 to 8 months at 12–18 °C and a relative humidity of 80–90%. During ripening, cheese rind is treated with oil to inhibit molds. The ripening is related mainly to proteolysis by the raw milk thermophilic lactic acid bacteria and milk plasmin, also residual rennet activity may be involved, while lipolysis is due to the use of rennet paste.

   The cheese shape is characteristic with square rectangular section (one side of ca. 15–18 cm and the other side of 43–53 cm) with blunt corners, and the possible

Fig. 6.193 Provolone Valpadana cheesemaking. Curd ripening. The photo has been kindly supplied by the Consorzio Provolone Valpadana PDO

presence of prints of the strings used to hang the cheeses during ripening. The weight ranges between 10 and 16 kg and the texture is hard or semi-hard. The cheese paste is usually white or slightly straw-colored and may become darker after a long ripening period, its structure is compact, elastic, and uniform, although the presence of little holes is possible, as is the presence of a few openings (slots) in long-ripened cheeses. The rind is thin and smooth, slightly straw-colored or yellow in young cheese that may become strawberry or brown after long ripening. The rind may be treated with olive oil and the cheese may be smoked. The flavor of the young cheeses is sweet and delicate but becomes spicy for long-ripened cheeses due to the action of the rennet paste lipase.

Fig. 6.194   Provolone del Monaco cheese. The photo has been kindly supplied by ONAF (Italian Cheese Taster Organisation, www.onaf.it)

6.4.5.4   Caciocavallo Silano

Caciocavallo Silano is a PDO semihard *pasta filata* cheese, with cooked paste, produced exclusively from raw cow's milk ripened for at least 30 days (Mipaaf 2017; Mucchetti et al. 2017; AFIDOP 2016; Esposito et al. 2014; Manzi et al. 2007; Mucchetti and Neviani 2006; Coppola et al. 2003; Villani et al. 1991; Grazia et al. 1990; De Caprariis 1912; Besana 1911) (Fig. 6.198 and 6.199). Manufacture is limited to some territories of Calabria, Campania, Molise, Puglia, and Basilicata regions. The large area of the typical territory of production and the abundant number of little dairies, partially justify the multiplicity of different protocols for making this cheese.

Milk may be stored at 4 °C after milking. Raw milk or thermized milk (58 °C per 30 s) may be used for cheesemaking. A natural whey culture is added to the vat milk. Lamb and/or kid rennet paste is used for coagulation at 36–38 °C for 13–50 min. The coagulum is broken into granules the size a hazel nut. Then, the curd is cooked by adding warm whey at a temperature close to 80 °C. The curd is held under warm diluted whey to favor acidification necessary for stretching. The broken curd may also be placed on tables at a controlled temperature to complete whey drainage and curd acidification. Under these conditions, a deep acidification of the curd by thermophilic lactic acid bacteria strains occurs. The curd is ready for stretching when it has reached pH 5.4–4.9. Sometimes, the right moment for stretching is based on empirical but skilled assessment. The different characteristics of the curd can contribute significantly to the loss of fat in the stretching water. The curd is cut into thin slices, placed in a container and covered with warm water at ca. 80–85 °C. The *pasta filata* is then formed manually and the curd cooled using cold

Fig. 6.195 Geographical area for the manufacture of Provolone del Monaco cheese

water. Curd is usually salted in brine (16–18°Bé) for 2–3 days. After salting, the curd is bound in pairs with plant or plastic ropes, and is then placed to dry in a warm room for 1–2 days in order to exude some of the cheese fat and give the crust an intense straw-yellow color. At this stage, Caciocavallo may be smoked. Ripening time ranges between a minimum of 30 days and 4–8 months.

The cheese has a characteristic truncated conical or an oval shape. The weight ranges between 1 and 5 kg, and the texture is hard or semi-hard. The cheese paste is usually white or slightly straw-colored and its structure is compact. In some cases, a limited number of holes appear close to the center of the cheese, probably because of the delay of NaCl diffusion. The rind is thin. The cheese taste is aromatic, delicate and with a tendency to sweet when young, intense and spicy when aged due to the proteolysis and lipolysis caused by the use of lamb or kid rennet paste.

Fig. 6.196 Ragusano cheese. The photo has been kindly supplied by ONAF (Italian Cheese Taster Organisation, www.onaf.it)

6.5 Goat's Milk Cheeses

Goat's milk cheese is probably one of the oldest traditions of dairy farming in the world. Goat breeding is usually characteristic of geographic areas where bovine breeding is difficult. In Italy, the manufacture of goat's milk cheeses has a local character and in many cases cheesemaking takes place at farm level. For the same reason, in most cases milk is used whole and raw. This implies a very large variety of goat's milk cheeses but it is extremely difficult to describe standardized protocols for manufacture (Mucchetti and Neviani 2006).

At the industrial level, transformation is limited to fresh acid-coagulated cheeses (formaggi caprini). The consumer's attention to fresh produce and high yields (thanks to their water content, often over 70%) have favored their production at an industrial level. This category differs from all other cheeses since acidification of the curd is not parallel or subsequent to serum purification but precedes it. Acid-coagulated cheeses are mostly consumed fresh, but in the version with mold growth and diffusion over the cheese rind they can have an extensive range of tastes and aromas.

In 2016, the milk used for Italian production of goat's cheese has been estimated to be close to 21,500 tonnes (Assolatte 2016). Cheeses made from goat's milk blended with sheep's milk are different (see descriptions above).

Fig. 6.197 Geographical area for the manufacture of Ragusano cheese

6.5.1 *Formaggella di Luinese*

Formaggella di Luinese is a PDO semi-hard cheese produced with whole raw goat's milk (Mipaaf 2017). The territories for goat raising, milk production, and cheese-making are in the northern area of the province of Varese (Lombardy region), also known as *Varesine Prealpi* (Figs. 6.200 and 6.201).

The raw goat's milk may be stored for a maximum of 30 h at a temperature less than 4 °C before cheesemaking. Natural milk or whey cultures or commercial/selected starters may be used. The main microbial composition is thermophilic lactic acid bacteria, and in some cases, mesophilic strains are added. The rennet must be exclusively natural calf rennet. Renneting occurs at 32–34 °C in 30–40 min. Breaking the coagulum takes place when the texture is consistent and proceeds until particles are of maize grain size. If the cheesemaking environment is particularly

Fig. 6.198 Caciocavallo Silano cheese. The photo has been kindly supplied by AFIDOP, Italy

cold (e.g., during winter), the temperature is increased to a maximum of 38 °C, and stirring follows for ca. 15 min. Molding takes place and the whey drainage is extended to a maximum of 48 h at room temperature, during which the cheese is turned at least 2–5 times. Salting is done using dry salt or brine (16–18°Bé), and drying at room temperature follows. Usually, ripening is in cells with a humidity of 85–95% and a temperature of ca. 15 °C. Ripening lasts for at least 20 days. The cheese microbiota depends mainly on autochthonous raw milk strains and on natural cultures when used. The ripened cheese has a dry matter content higher than 45%, and a fat content higher than 41% of cheese dry matter.

The cheese is cylindrical in shape, 4–6 cm high and 13–15 cm in diameter, and weighs 700–900 g. The cheese paste is soft, elastic, and compact, in some cases with a few small holes. The color of the cheese paste is homogeneous and mainly white. Usually, molds are present on the cheese rind. The taste of cheese is sweet, delicate, and pleasant, and it intensifies with the extent of ripening. The smell and flavor are delicate, also becoming more intense during ripening.

6.6   Dairy Products Obtained by Thermocoagulation of Milk, Cream, or Whey

A particular category of Italian dairy products, consisting mainly of Ricotta and Mascarpone, is obtained by thermocoagulation of milk proteins, in the absence of a starter microbiota and rennet activity. These products, served and consumed like cheeses, also belong to the Italian tradition and are very popular (Mucchetti and Neviani 2006).

Fig. 6.199 Geographical area for the manufacture of Caciocavallo Silano cheese

6.6.1   *Mascarpone*

Mascarpone is a rindless, soft, spreadable and unripened dairy product, obtained by the acid-heat coagulation of milk cream, to which different amounts of milk may sometimes be added (Fig. 6.202). It is produced without any bacterial fermentation. Its manufacture is spread throughout Italy without particular reference to some region or province (Mucchetti and Neviani 2006; Carminati et al. 2000, 2001; Franciosa et al. 1999; Battelli et al. 1995; Resmini et al. 1984).

The main component of Mascarpone is milk fat (usually higher than 40%). When a light variety is produced (with fat less than 35%), the amount of protein (casein and whey proteins) required for keeping a correct structure is increased. Traditionally,

Fig. 6.200 Formaggella di Luinese cheese. The photo has been kindly supplied by ONAF (Italian Cheese Taster Organisation, www.onaf.it)

it is produced from pasteurized cream. The traditional protocol involves preparation of the cream, obtained by centrifugal separation or by spontaneous flotation, with a fat content of 25–40%. Then, the cream is heated in small pots, immersed in a warm water bath to reach 80–90 °C, to which is added a solution of acetic or citric acids. The addition of acid may vary depending on the inherent acidity of the cream. Whey is usually drained off using centrifugal separation. Nowadays, ultrafiltration may be used for whey drainage. Industrial production involves some innovations. In some cases, the cream is Ultra High Temperature (UHT) treated, and it is possibly mixed with milk concentrate prepared by ultrafiltration. Different ingredients may be used to reduce the pH of the cream but usually citric acid is preferred. After coagulation, the cream is usually hot-filled into packs and then cooled to 4 °C, when the final structure develops. The shelf-life is usually up to 60 days. The temperature of manufacture, a_w and the pH do not guarantee the destruction or the inhibition of the germination of *Clostridium botulinum* spores, so this dairy product must be kept refrigerated (4 °C).

Mascarpone is typically white or, when cows are fed on grass, light pale in color. It has the mild sweet taste of cream and fresh butter-like flavor; its consistency is smooth and spreadable, also at low temperature. Mascarpone can be used as an ingredient in other foods, especially desserts.

6.6.2   Ricotta

La Ricotta is an original, typically Italian dairy product, even if some variants are produced in neighboring countries, to enhance the value of residual whey from cheesemaking (Fig. 6.203). Despite the importance of Ricotta production, its quantification escapes official statistics. The difficulty in quantifying the national

Fig. 6.201 Geographical area for the manufacture of Formaggella di Luinese cheese

production of Ricotta is due to the extreme fragmentation of small producers. In the South of Italy and in insular Italy about 30 different types of Ricotta could be classified, based on the origin of the raw material (cow's milk, sheep's milk, buffalo's milk, goat's milk), production technique or conservation technique (fresh, salted, bake, lemon, etc.). The production of such Ricotta types, especially those obtained from non-bovine whey, is often seasonal and is linked in large part to the agropastoral tradition, characterized by a strong dispersion in the territory. Similarly, in the mountain pastures or in the self-processing companies of the bottom-valley milk located in the mountain areas of the Alpine and pre-Alpine ranges of Northern Italy, the production of Ricotta is very variable due to both seasonal and contingent product requests (Mucchetti and Neviani 2006; Contarini et al. 2002).

Ricotta production is the most widespread way for using cheese whey. Ricotta may be defined as a rindless soft fresh dairy product, with a grainy appearance but

Fig. 6.202 Mascarpone. The photo has been kindly supplied by Dr. Elena Bancalari (University Parma, Italy)

Fig. 6.203 Ricotta. The photo has been kindly supplied by Dr. Elena Bancalari (University Parma, Italy)

smooth to the taste, and spreadable. Ricotta belongs to the group of whey dairy products obtained by the coagulation of whey proteins by heat with or without the addition of lactic or citric acids, or calcium and/or magnesium salts to modify the ionic strength.

It may be produced using whey and other milk ingredients obtained from cow, ewe, water buffalo, or goat milk or their blends. Whole milk, skim milk, milk cream or whey cream, and sometimes NaCl, may be added to the whey. It is usually a fresh product but it could be ripened or smoked. A large number of different variants of Ricotta are produced and some of them may be semi-soft or even hard, depending on the type of salting, ripening and smoking processes.

Fresh Ricotta is characterized by a high content of water, with a delicately sweet taste of milk and cream, and a granular but non-sandy texture. Its color is white, with different tones according to the animal species of origin of the raw materials. The addition of different amounts of milk or cream to whey changes the color and structure of Ricotta. The composition is very variable because of the different characteristics and proportions of the dairy ingredients used (Contarini et al. 2002). The origin of whey (sheep, buffalo, cow and goat's milk) and the amount of milk added, influence the protein content of the blend, as well as the addition of whey powders or milk proteins.

Ripened Ricotta is a product with a delicately spicy taste and a moisture content of 50–55%. In Italy, the denomination of two products, Ricotta Romana and Ricotta di Bufala Campana, is protected by the PDO system. Ricotta is used also as a food ingredient.

6.6.3 Ricotta Forte

Ricotta Forte is a long-ripened variety of Ricotta manufactured from ewe's milk. The manufacture is typical for the regions of the Southern of Italy, in particular the Apulia region.

The whey from ewe's milk cheesemaking is subjected to thermo-coagulation at more than 90 °C with or without the addition of lactic or citric acids, which favor the precipitation of whey proteins. After precipitation and cooling, a small amount of salt is added (the final concentration in the product is less than 1%). Ripening takes place at room temperature in a vat with the capacity of ca. 1000 L. No starters are added. Ripening lasts 10–12 months, and daily the mass is turned/mixed to prevent fungal growth in the surface. The biochemical activities, proteolysis and mainly lipolysis, during ripening are extremely intense and rely on the whey autochthonous and house microbiota, consisting of bacteria, yeasts and molds.

This dairy product is a cream with a strongly aromatic flavor and spicy taste. It is used mainly as a flavoring agent for many dishes, and is rarely consumed alone.

References

Addeo F and Laezza P (1987) An electrophoretic study of casein proteolysis in the Provolone cheese. In: atti "CNR-IPRA Third Subproject: conservation and processing of foods." Abstract 161, pp 492–495

Addeo F, Mucchetti G, Neviani E (1997a) Gli aspetti biochimici della maturazione del formaggio con particolare riferimento alle varietà a pasta dura. Scienza Tecnica Lattiero Casearia 48:7–15

Addeo F, Chianese L, Masi P et al (1997b) La Mozzarella: un formaggio tradizionale in evoluzione. Ed Mofin, Novara

Addis M, Fiori M, Riu G et al (2015) Physico-chemical characteristics and acid profile of PDO Pecorino Romano cheese: seasonal variation. Small Rum Res 126:73–79

AFIDOP-IGP (2016) FIL-IDF Italy and AFIDOP-IGP 2016 Raccolta bibliografica dei parametri tecnologici e compositivi dei formaggi DOP

Albenzio M, Corbo MR, Rehman SU et al (2001) Microbiologcal and biochemical characteristics of Canestrato Pugliese cheese made from raw milk, pasteurized milk or by heating the curd in hot whey. Int J Food Microbiol 67:35–41

Arnaudi C (1948) L'impiego di penicilli selezionati nella fabbricazione del Gorgonzola. Mondo Latte 2:3–6

Assolatte (2016) Italian Dairy Association—report

Avellini P, Clementi F, Trabalza Marinucci M et al (1999) "Pit" cheese: compositional, microbiological and sensory characteristics. Ital J Food Sci 11:317–333

Barzaghi S, Davoli E, Rampilli M et al (1997) La lipolisi nel formaggio Provolone: ruolo del caglio in pasta. Scienza Tecnica Lattiero Casearia 48:146–156

Battelli G, Pellegrino L, Ostini G (1995) Alcuni aspetti della qualità della panna e del Mascarpone industriale da essa ottenuto. Latte 20:1098–1112

Battistotti B, Bottazzi V, Vola G (1976) Ricerche sulle caratteristiche microbiologiche e chimiche del formaggio Fontina. Scienza Tecnica Lattiero Casearia 27:29–45

Battistotti B, Bosi F, Bottazzi V (1980) Ricerche sul sieroinnesto per Provolone. I parte: aspetti tecnologici e microbiologici. Industria Latte 26:19–26

Battistotti B, Bottazzi V, Gonzaga E et al (1986) Fermentazione lattica e demineralizzazione della pasta per Provolone. Scienza Tecnica Lattiero Casearia 37:117–123

Berard J, Bianchi F, Careri M et al (2007) Characterization of the volatile fraction and free fatty acids of "Fontina Valle d'Aosta", a protected designation of origin Italian cheese. Food Chem 105:293–300

Bernini V, Dalzini E, Lazzi C et al (2016) Cutting procedures might be responsible for *Listeria monocytogenes* contamination of foods: the case of Gorgonzola cheese. Food Control 61:54–61

Bertolino M, Dolci P, Giordano M et al (2011) Evolution of chimico-physical characteristics durino manufacture and ripening of Castelmagno PDO cheese in wintertime. Food Chem 129:1001–1011

Besana C (1886) Le Fontine di Val d'Aosta. Annuario Regia Stazione Sperimentale di Caseificio in Lodi Anno 1886. Lodi, Tipografia Costantino dell'Avo, pp 98–105

Besana C (1911) Il caseificio meridionale ed il suo avvenire. Industria Lattiera Zootecnica 9:157–158, 173–174, 189–190, 206–207, 220–221

Besana C (1915) Sulla fabbricazione di formaggi a pasta molle d'inverno. Industria lattiera e Zootecnica 13:3–4

Bianchi Salvatori B, Sacco M (1981) Indagine sulla tecnologia e maturazione del formaggio Silter. Industria Latte 17:3–30

Bizzarro R, Neviani E, Mucchetti G et al (1999) Il Formaggio Pecorino Toscano. Quaderno ARSIA

Bizzarro R, Torri Tarelli G et al (2000) Phenotypic and genotypic characterization of lactic acid bacteria isolated from Pecorino Toscano Cheese. Ital J Food Sci 12:303–316

Bodini F (1973) Aspetti della tecnologia del formaggio Provolone. Scienza Tecnica Lattiero Casearia 24:229–235

Bontempo L, Larcher R, Camin F et al (2011) Elemental and isotopic characterization of typical Italian alpine cheeses. Int Dairy J 21:441–446

Bottazzi V, Battistotti B (1973) Aspetti scientifici e tecnici della produzione del Provolone. Scienza Tecnica Lattiero Casearia 24:216–228

Bottazzi V, Arrizza S, Ledda A (1978) Impiego di colture di fermenti lattici nella produzione di Fiore Sardo. Scienza Tecnica Lattiero Casearia 29:160–168

Braidot S (1948) Il formaggio Montasio. Latte 22:8–12

Candido A (1994) Il Pecorino Siciliano. II. Aspetti microbiologici e compositivi dei prodotti freschi (tuma e primosale). Scienza Tecnica Lattiero Casearia 45:249–256

Cantoni C, Chiesa LC, Cesana F (2005) Analisi delle sostanze volatili del Taleggio. Industrie Alimentari 451:971–977

Carbone E (1957) Ricerche sulla composizione del formaggio Fontina. Rivista del Latte 13:1–9

Carini S (1990) Ruolo degli enzimi nella maturazione del Gorgonzola. Latte 15:383–388

Carini S, Baglieri G (1981) Preparazione ed impiego di uno starter nella produzione di Caciocavallo Ragusano. Industria Latte 17:57–69

Carini S, Galli A, Ottogalli G et al (1969) Valutazione della composizione microbiologica ed azotata del formaggio Taleggio durante la maturazione. Scienza Tecnica Lattiero Casearia 20:63–94

Carini S, Francani R, Toppino PM (1985) Criteri di tipizzazione del Provolone; analisi chimiche di base e degli additivi utilizzati. Annali Istituto Sperimentale Lattiero Caseario di Lodi, pp 77–98

Carminati D, Grossi S, Somenzi D et al (2000) Indagine sulla presenza di spore di botulino in panne di centrifuga ed affioramento. Industrie Alimentari 39:996–1001

Carminati D, Perrone A, Neviani E (2001) Inhibition of *Clostridium sporogenes* growth in Mascarpone cheese by co-inoculation with *Streptococcus thermophilus* under conditions of temperature abuse. Food Microbiol 18:571–579

Carminati D, Mucchetti G, Neviani E et al (2004) Il formaggio Gorgonzola ed il problema *Listeria monocytogenes*. Quaderni della Ricerca della Regione Lombardia 28:1–44

Cecchi L, Deiana P, Catzeddu P et al (1998) Studies on lactic acid isomers in Pecorino Romano cheese by means of radiotracers. Milchwissenschaft 53:84–88

Coisson JD, Arlorio M, Martelli A (2000) Caratterizzazione chimica del formaggio Robiola di Roccaverano DOP. Scienza Tecnica Lattiero Casearia 51:38–49

Contardi R (1965) Tecnologia del formaggio Pecorino Romano. Latte 39:196–201

Contarini G, Toppino PM (1995) Lipolysis in Gorgonzola cheese durino ripening. Int Dairy J 5:141–145

Contarini G, Povolo M, Avalli A et al (2002) Minor components of lipid fraction of cow's, ewe's, goat's and buffalo's whey cheese (ricotta). In: Atti di Congrilait 2002, "XXVI International Dairy Congress" Parigi, 24–27 settembre 2002, Poster

Coppola S, Parente E, Dumontet S et al (1988) The microflora of natural whey cultures utilized as starters in the manufacture of Mozzarella cheese from water-buffalo milk. Lait 68:295–310

Coppola R, Succi MA, Sorrentino E et al (2003) Survey of lactic acid bacteria during the ripening of Caciocavallo cheese produced in Molise. Lait 83:211–222

Corbo MR, Albenzio M, De Angelis M et al (2001) Microbiologcal and biochemical properties of Canestrato Pugliese hard cheese made supplemented with bifidobacteria. J Dairy Sci 84:551–561

Corradini C, Battistotti B, Fenaroli A (1973) Ricerche sul formaggio Silter. Latte 47:168–173

Corsetti A, Rossi J, Gobbetti M (2001) Interactions between yeasts and bacteria in the smear surface ripened cheeses. Int J Food Microbiol 69:1–10

Cozzi G, Ferlito J, Pasini G et al (2009) Application of near-infrared Spectroscopy to chemical and color analysis to desciminate the production chains of Asiago d'allevo cheese. J Agric Food Chem 57:11449–11454

Dalla Torre G, Fontanella E (1955) Composizione chimica e valore alimentare di alcuni formaggi tipici del Veneto. Latte 29:697–699

De Caprariis T (1912) Tipi di formaggi meridionali - Il Caciocavallo. Industria Lattiera Zootecnica 10:154–155

De Dea Lindner J, Bernini V, De Lorentiis A et al (2008) Parmigiano Reggiano cheese: evolution of cultivable and total lactic microflora and peptidase activities during manufacture and ripening. Dairy Sci Technol 88:511–523

De Angelis M, Gobbetti M (2011) Traditional pasta-filata cheese. In: Fuquay JW, Fox PF, McSweeney PLH (eds) Encyclopedia of dairy sciences, vol. 1, 2nd edn. Academic, San Diego, pp 745–752

De Pasquale I, Di Cagno R, Buchin S et al (in press) Effect of selected autochthonous nonstarter lactic acid bacteria as adjunct cuktures for making Pecorino Crotonese cheese. Int J Food Microbiol

Deiana P, Fatichenti F, Farris GA et al (1984) Metabolization of lactic and acetic acids in Pecorino Romano cheese made with a combined starter of lactic acid bacteria and yeasts. Lait 64:380–394

Deiana P, Rossi J, Caredda M et al (1997) La microflora secondaria nel Pecorino Romano. Scienza Tecnica Lattiero Casearia 48(6):487–500

Delforno G (1958) Il formaggio Bra. Istituto Zootecnico Caseario per il Piemonte, Tipografia Antonio Cordani Spa Milano

Delforno G (1967) Tecnica di fabbricazione del formaggio Taleggio. Industria Latte 3:227–233

Delforno G (1970) Tecnologia del formaggio Montasio. Mondo del Latte 24:509–514

Delforno G (1975) Il formaggio Toma. Mondo del Latte 29:641–647

Delforno G (1982) Il formaggio Pannerone. Mondo del Latte 36:431–433

Delforno G, Fondrini A (1966) Il formaggio Branzi. Mondo latte 20:864–868

Delforno G, Losi A (1975) Il formaggio Silter. Mondo Latte 29:20–21

Di Bidino R (1979) Il formaggio Montasio. Latte 4:1069–1071

Di Cagno R, De Pasquale I, De Angelis M et al (2011) Manufacture of Italian Caciotta-type cheeses with adjuncts and attenuated adjuncts of selected non-starter lactobacilli. Int Dairy J 21:254–260

DOOR (2017) DOOR: EU-database of agricultural products and foods

Emaldi GC (1960) Il Provolone. Rivista del Latte 24:3–11

Ercolini D, Moschetti G, Blaiotta G et al (2001) The potential of a polyphasic PCR-DGGE approach in evaluating microbial diversità of natural whey cultures for water-buffalo Mozzarella cheese production: bias of culture-dependent and culture-independent analyses. Syst Appl Microbiol 24:610–617

Esposito G, Masucci F, Napolitano F et al (2014) Fatty acid and sensorory profiles of Caciocavallo cheese as affected by management system. J Dairy Sci 97:1918–1928

Fascetti G (1903) Il formaggio Bra. Industria del Latte 1:97–99

Fascetti G (1935) Enciclopedia del caseificio. Hoepli Editore, Milano

Fascetti G, Savini E (1929) Ricerche sul formaggio Pecorino Romano (Terza Comunicazione). Annali Istituto Sperimentale di Caseificio di Lodi (Fascicolo 4–5)

Fazzino S, Randazzo CL, Torriani S (2002) Evoluzione e caratterizzazione della microflora lattica autoctona del formaggio Ragusano. Industrie Alimentari 41:290–294

Fox P, Guine T, Cogan T et al (eds) (2017) Fundamental of cheese science, 2nd edn. Springer, New York

Franciosa G, Pourshaban M, Gianfranceschi M et al (1999) *Clostridium botulinum* spores and toxin in Mascarpone cheese and other milk products. J Food Prot 62:867–871

Fulco A, Candido A, Apollo S et al (1984) Rilievi sulla tecnologia di produzione del Ragusano. Industria Latte 20:3–18

Gatti M, Lazzi C, Rossetti L et al (2003) Biodiversity in *Lactobacillus helveticus* strains in natural whey starter used for Parmigiano Reggiano Cheese. J Appl Microbiol 95:463–480

Gatti M, De Dea Lindner J, Turroni F et al (2008a) Microbiological and proteolytic aspects of Parmigiano Reggiano cheese ripening. In 5th IDF Symposium on cheese ripening, Berna 9–13 March 2008

Gatti M, De Dea Lindner J, Gardini F et al (2008b) A model to assess microbial enzyme activities in Parmigiano Reggiano. J Dairy Sci 91:4129–4137

Gatti M, De Dea Lindner J, De Lorentiis A et al (2008c) Dynamics of entire and lysed bacterial cells during Parmigiano Reggiano cheese production and ripening. Appl Environ Microbiol 74:6161–6167

Gatti M, Bottari B, Lazzi C et al (2014) Microbial evolution in raw milk, long-ripened cheeses produced using undefined natural whey starters. Invited review. J Dairy Sci 97:573–591

Ghitti C (1959) Tecnologia dei formaggi di pasta filata a maturazione media del tipo "Provolone". Latte 33:388–389, 487–488, 491–492

Ghitti C, Bianchi-Salvadori B (1985) Il formaggio Italico. Collana Tecnologica Centro Sperimentale del Latte, CSL, Milano

Ghitti C, Ottogalli G (1987) Il formaggio Crescenza. Collana Tecnologica Centro Sperimentale del Latte, CSL, Milano

Giraffa G, Neviani E (1999) Different *Lactobacillus helveticus* strain popolations dominate during Grana Padano cheese-making. Food Microbiol 16:205–214

Giraffa G, Rossetti L, Mucchetti G et al (1998) Influence of the temperature gradient on the growth of thermophilic lactobacilli used as natural starter in Grana cheese. J Dairy Sci 81:31–36

Gobbetti M (2004) Extra-hard varieties. In: Fox PF, McSweeney PLH, Cogan TM et al (eds) Cheese: chemistry, physics and microbiology. Elsevier Ltd, London, pp 51–61

Gobbetti M, Di Cagno R (2002) Hard Italian cheeses. In: Roginski H, Fox PF, Fuquay JW (eds) Encyclopedia of dairy sciences. Academic, London, pp 378–385

Gobbetti M, Di Cagno R (2017) Extra-hard varieties. In: McSweeney PLH, Fox PF, Cotter PD, Everett DW (eds) Cheese chemistry, physics and microbiology, 4th Academic London 2 p 809–828

Gobbetti M, Burzigotti R, Smacchi E et al (1997a) Microbiology and biochemistry of Gorgonzola cheese during ripening. Int Dairy J 7:519–529

Gobbetti M, Lowney S, Smacchi E et al (1997b) Microbiology and biochemistry of Taleggio cheese during ripening. Int Dairy J 7:509–517

Gobbetti M, Folkertsma B, Fox PF et al (1999) Microbiology and biochemistry of Fossa (pit) cheese. Int Dairy J 9:763–773

Gorini C (1936) Diffondiamo le latterie-scuola alpine. Latte 10:9–20

Grassi MA, Civera T, Pattono D et al (2002) Caratteristiche microbiologiche della Robiola di Roccaverano. Industrie Alimentari 41:321–1327

Grazia L, Grisafi G, Tini V (1990) Evoluzione della microflora batterica nella lavorazione della Provola Silvana. Industrie Agrarie 31:67–70

Guerriero P, Rottigni C, Todaro A et al (1997) Influenza della tipologia dello starter sulle caratteristiche della pasta e della maturazione del formaggio Crescenza: indagine preliminare. Industria Latte 33:33–48

Gusmeroli F, Pino G, Bottini F et al (1988) Il formaggio Casera di Valtellina. Indagine sulla tecnologia di produzione. Industria Latte 24:55–68

Innocente N (1997) Free amino acids and water soluble nitrogen as ripening indices in Montasio cheese. Lait 77:359–369

La Curlo S (1925) Pecore e formaggi pecorini in Sicilia. Industria Lattiera Zootecnica 23:59–61

Ledda A, Arrizza S (1969) Rilievi sulla tecnologia di produzione del formaggio Pecorino "Romano" in Sardegna. Scienza Tecnica Lattiero Casearia 20:143–183

Ledda A, Murgia A, Arrizza S (1978) Indagine sulla tecnologia di fabbricazione del Fiore Sardo. Scienza Tecnica Lattiero Casearia 29:145–159

Ledda A, Scintu MF, Pirisi A et al (1994) Caratterizzazione tecnologica di ceppi di lattococchi e di enterococchi per la produzione di formaggio pecorino Fiore Sardo. Scienza Tecnica Lattiero Casearia 45:443–456

Licitra G (2001) Ragusano cheese: a work of art. Caseus Int 1:18–32

Licitra G, Portelli G, Campo P et al (1998) Technoloy to produce Ragusano cheese: a survey. J Dairy Sci 81:3343–3349

Licitra G, Campo P, Vanenti M et al (2000) Composition of Ragusano cheese during aging. J Dairy Sci 83:404–411

Maggi E, Ballerini P (1994) Il formaggio Monte Veronese. Industrie Alimentari 33:390–392

Malacarne M, Summer A, Panari G et al (2006) Caratterizzazione chimico-fisica della maturazione del Parmigiano-Reggiano. Scienza e Tecnica Lattiero-Casearia 57:215–228

Mangia NP, Murgia MA, Garau G et al (2008) Influence of selected lab cultures on the evolution of free amino acids, free fatty acids, and Fiore Sardo cheese microflora during the ripening. Food Microbiol 25:366–377

Mangia NP, Murgia M, Garau G et al (2011) Microbiological and physicochemical properties of Pecorino Romano cheese produced using a selected starter culture. J Agric Sci Technol 13:585–600

Mangia NP, Murgia MA, Garau G et al (2013) Suitability of selected autochthonous lactic acid bacteriacultures for Pecorino Sardo Dolce cheese manufacturing: influence on microbial composition, nutritional value and sensory attributes. Int J Dairy Technol 66:543–551

Mannu L, Paba A (2002) Genetic diversity of lactococci and enterococci isolated from home-made Pecorino Sardo ewe's milk cheese. J Appl Microbiol 92:55–62

Mantovani R, Bailoni L, Chatel A et al (2003) Relationship between pasture and nutritional aspects of Fontina cheese manufacture in alpine farms. Ital J Animal Sci 2:287–289

Manzi P, Pizzoferrato L (2009) Kinetic study on unsaponificable fraction changes and lactose hydrolysis during storage of Mozzarella di Bufala Campana PDO cheese. Int J Food Sci Nutr 60:1–10

Manzi P, Marconi S, Di Costanzo MG et al (2007) Composizione di formaggi DOP italiani. La Rivista di Scienze dell'Alimentazione 36:9–22

Marcialis A, Pettinau M, Bottazzi V (1968) Composizione chimica del formaggio Pecorino "Romano" prodotto in Sardegna. Scienza Tecnica Lattiero Casearia 19:411–422

Martelli A (1989) Componenti volatili dell'aroma del formaggio Gorgonzola. Rivista Italiana Scienza dell'Alimentazione 18:251–262

Massa S, Turtura GC, Trovatelli LD (1988) Qualité hygiénique du fromage de "fosse" de Sogliano al Rubicone (Italie). Lait 68:323–332

Mauriello G, Moio L, Moschetti G et al (2001) Characterization of lactic acid bacteria strains on the basis of neutral compounds produced in whey. J Appl Microbiol 90:928–941

Mauriello G, Moio L, Genovese A et al (2003) Relationships between flavoring capabilities, bacterial composition ang geographical origin of natural whey cultures used for traditional water-buffalo Mozzarella cheese manufacture. J Dairy Sci 86:486–497

Merlo B (2001) Il consorzio tutela Valtellina Casera e Bitto. Latte 26:43–64

Merlo B (2002) Il Consorzio Tutela formaggio Asiago. Latte 27:63–78

Migliorisi D, Caggia C, Pulvirenti A et al (1997) I batteri lattici del Pecorino Siciliano. Latte 22:80–86

Minervini F, Conte A, Del Nobile MA et al (2017) Dietary fibers and protective lactobacilli drive burrata cheese microbiome. Appl Environ Microbiol 83. https://doi.org/10.1128/AEM.01494-17

Mipaaf (2017) www.Mipaaf—formaggi DOP. https://www.politicheagricole.it

Moio L, Dekimpe J, Etievant PX et al (1993) Comparison of the neutral volatile compounds in Mozzarella cheese made from bovine and water buffalo milk. Ital J Food Sci 5:215–225

Moio L, Piombino P, Addeo F (2000) Odour-impact compounds of Gorgonzola cheese. J Dairy Res 67:273–285

Morandi S, Brasca M, Lodi R (2011) Technological, phenotypic and genotypic characterisation of wild lactic acid bacteria involved in the production of Bitto PDO Italian cheese. Dairy Sci Technol 91:341–259

Mucchetti G, Neviani E (eds) (2006) Microbiologia e Tecnologia Lattiero-Casearia. Tecniche Nuove, Milano

Mucchetti G, Neviani E, Carminati D et al (1985) Il formaggio "nostrano" di Valle Trompia: indagine tecnologica e composizione chimica. Industria Latte 21:23–40

Mucchetti G, Taglietti P, Emaldi GC et al (1993) Influenza di alcuni fattori tecnologici ed ambientali sulle caratteristiche organolettiche del formaggio di Valle Trompia. Latte 18:548–556

Mucchetti G, Carminati D, Addeo F (1997) Tradition and innovation in the manufacture of the water buffalo Mozzarella cheese produced in Campania. In: Proceedings 5th World Buffalo Congress, Caserta 13–16 ottobre 1997, pp 173–181

Mucchetti G, Addeo F, Neviani E (1998) Evoluzione storica della produzione di formaggi a Denominazione di origine Protetta (DOP). 1. Pratiche di produzione, utilizzo e composizione dei sieroinnesti nella caseificazione a formaggi Grana Padano e Parmigiano-Reggiano: considerazioni sulle relazioni tra sieroinnesto e DOP. Scienza Tecnica Lattiero casearia 49:281–311

Mucchetti G, Bonvini B, Remagni MC et al (2008) Influence of cheese-making technology on composition and microbiological characteristics of Vastedda cheese. Food Control 19:119–125

Mucchetti G, Pugliese A, Paciulli M (2017) Characteristics of some important Itaian cheeses—Parmigiano Reggiano, Grana Padano, Mozzarella, Mascarpone and Ricotta. In: Cruz RMS, Veira MC (eds) Mediterranean foods. CRT Press, London

Neviani E, Carini S (1994) Microbiology of Parmesan cheese. Microbiol Alim Nutr 12:1–8

Neviani E, Gatti M (2013) Microbial evolution in raw-milk, long-ripened cheeses: Grana Padano and Parmigiano Reggiano case. In: Randazzo CL, Caggia C, Neviani E (eds) Cheese ripening: quality, safety and health aspects. Nova Science Publishers, Hauppauge, pp 133–148

Neviani E, Forni F, Bossi MG et al (1992) Il siero-innesto per Provolone: specie dominanti e loro attività biochimiche. Scienza Tecnica Lattiero Casearia 43:229–240

Neviani E, Bizzarro R, Righini A et al (1998) Pecorino Toscano DOP: tecniche di produzione e caratteristiche microbiologiche. Industria Latte 34:3–35

Ottogalli G, Galli A, Rondinini G et al (1975) La microbiologia del formaggio Pannerone. Industria Latte 11:7–17

Ottogalli G, Carminati D, Franzetti L et al (1996) Surface bacterial microflora of Taleggio cheese. Microbiol Alim Nutr 14:39–42

Paleari MA, Beretta G, Tiraboschi M (1991) Un prodotto tipico dell'alta Val Brembana: il "Formai de Mut". Scienza Tecnica Lattiero Casearia 42:286–289

Paleari MA, Soncini G, Beretta G et al (1993) Microbiologia e qualità del formaggio "Semigrasso di Monte" della Valsassina. Industria Latte 29:69–85

Parisi O (ed) (1966) Il formaggio Grana, 4th edn. S.T.E.M. Mucchi, Modena

Patti A, Fulco A, Candido A et al (1984) Rilievi sulla tecnologia di produzione del Pecorino Siciliano. Latte 9:190–193

Patti A, Fulco A, Candido A et al (1985) Sul valore medio di alcuni parametri caratterizzanti il Pecorino Siciliano. Latte 10:114–119

Pattono D, Grassi MA, Civera T et al (2001) Profilo compositivo della Robiola di Roccaverano a carattere artigianale. Industrie Alimentari 40:1351–1355

Pavolotti G (1981) Il formaggio Raschiera. Mondo del Latte 35:686–688

Pecorari M, Gambini G, Panari G et al (2007) Caratterizzazione del Parmigiano-Reggiano di 12 mesi. Scienza e Tecnica Lattiero-Casearia 58:205–217

Pellegrino L, Hogenboom JA, Pazzaglia C et al (1995) La caratterizzazione analitica del formaggio Fontina sulla base della sua composizione in amminoacidi liberi. Latte 20:1086–1093

Pettinau M, Nuvoli G, Podda F (1978) Rilievi sulla composizione chimica del Fiore Sardo. Scienza Tecnica Lattiero Casearia 29:169–181

Pinarelli C (2005) La Casciotta d'Urbino. Il latte 9:90–92

Pinarelli C (2006) The Fontina Cheese. Il latte 31:50–52

Pirisi A, Pinna G, Addis M et al (2007) Relationship betwwen the enzymatic composition of lamb rennet paste and proteolytic, lipolytic pattern and texture of PDO Fiore Sardo ovine cheese. Int Dairy J 17:143–156

Pirrone L (1994) Il Pecorino Siciliano. La tecnologia di trasformazione. Scienza Tecnica Lattiero Casearia 45:199–210

Pisano MB, Fadda ME, Depilano M et al (2006) Microbiological and chemical characterization of Fiore Sardo, atraditional Sardinian cheese made from ewe's milk. Int J Dairy Technol 59:171–179

del Prato S (1998) Il formaggio "Crescenza". Latte 23:66–75

Randazzo CL, Torriani S, Akkermans ADL et al (2002) Diversità, dynamics, and activity of bacterial communities during production of an artisanal Sicilian cheese as evaluated by 16S rRNA analysys. Appl Environ Microbiol 68:1882–1892

Randazzo CL, Caggia C, Neviani E (eds) (2013) Cheese ripening: quality, safety and health aspects. Nova Science Publishers, Hauppauge

Renko P (1956) Importanza delle colture di fermenti selezionati per l'incremento della produzione e vendita del formaggio Taleggio. In: XIV Congresso Internazionale sul Latte e Derivati, Roma, Volume II, Parte II, Sezione II, Argomento, vol 4, pp 463–467

Resmini P, Pagani MA, Prati F (1984) L'ultrafiltrazione del latte nella tecnologia del Mascarpone. Scienza Tecnica Lattiero Casearia 35:213–230

Resmini PP, Pellegrino L, Hogenboom J et al (1987) Attempt to characterize Provolone cheese on a chemometric basis. In: CNR-IPRA Third Subproject: conservation and processing of foods Abstract 152, pp 469–472

Revello Chion A, Tabacco E, Giaccone D et al (2010) Variation of fatty acid and terpene profiles in mountain milk and "Toma Piemontese" cheese as affected by diet composition in different seasons. Food Chem 121:393–399

Romano R, Giordano A, Chianese L et al (2011) Tryacylglycerols, fatty acids and conjugated linoleic acids in Italian mozzarella di Bufala Campana cheese. J Food Compos Anal 24:244–249

Romanzin A, Corazzin M, Piasentier E et al (2013) Effect of rearing system (mountain pasture vs indoor) of Simmental cows on milk composition and Montasio cheese characteristics. J Dairy Res 80:390–399

Romanzin A, Corazzin M, Favotto S et al (2015) Montasio cheese liking as affected by information about cows breed and rearing system. J Dairy Res 82:15–21

Rossetti L, Fornasari ME, Gatti M et al (2008) Grana Padano cheese whey starters: Microbial composition and strain distribution. Int J Food Microbiol 127:168–171

Santarelli M, Gatti M, Lazzi C et al (2008) Whey starter for Grana Padano Cheese: effect of technological parameters on viability and composition of the microbial community. J Dairy Sci 91:883–891

Santoro M, Faccia M (1998) Influence of mold size and rennet on proteolysis and composition of Canestrato Pugliese cheese. Ital J Food Sci 10:217–228

Savini E (1945) La Crescenza. La Rivista del Latte 1:21–29

Savini E (ed) (1950) Il Pannerone. Seconda edizione. Istituto Sperimentale di Caseificio Lodi, Tipografia La Moderna, Lodi

Scapaccino M (ed) (1936) Il formaggio Asiago. S.A. Arte Stampa, Roma

Sforza S, Galaverna G, Neviani E et al (2004) Study of the oligopeptide fraction in Grana Padano and Parmigiano-Reggiano cheeses by liquid chromatography electrospray ionization mass spectrometry. Eur J Mass Spec 10:421–447

Spinelli G (1958) Tecnologia casearia: Quartirolo o stracchino quadro. Latte 32:581–585

Strozzi E (1941) Il "Fontina" della Val d'Aosta. Latte 15:6–18

Todesco R, Resmini P, Aquili G (1992) Indici chimico analitici del formaggio Crescenza correlabili alla sua struttura. Industria Latte 28:41–57

Tomarelli D (ed) (1951) Il formaggio pecorino. Manuale pratico per la produzione, la salagione e l'allevamento del formaggio pecorino, 2nd edn. Tipografia Agostiniana, Roma

Tosi E (1904) Le latterie della Carnia. Industria del Latte 82:90–91

Tripi V (1980) Indagine sulla composizione chimica del formaggio Pannerone. Mondo Latte 34:244–246

Villani F, Pepe O, Coppola R et al (1991) Aspetti microbiologici della fabbricazione del Caciocavallo Podolico. Latte 16:780–788

Zamorani A, Nicolosi AC, Cataldi Lupo MC (1974) Il Caciocavallo Ragusano. Composizione e maturazione, con particolare riguardo alla evoluzione delle sostanze azotate. Latte 48:1047–1105

Zamorani A, Spettoli P, Crapisi A et al (1987) Enzymatic typification of Provolone cheese. In: Atti CNR-IPRA, Third Subproject: conservation and processing of foods, pp 477–478

Zeppa G, Gerbi V, Tallone G (2002) Aspetti tecnologici, analitici e sensoriali. In: Il formaggio Ossolano. Supplemento al n 31 di "Quaderni della Regione Piemonte-Agricoltura", pp 215–244

Zucconi L (1874): Notizie sulla storia, fabbricazione e commercio dello Stracchino di Gorgonzola. Atti e Memorie del Primo Congresso per l'incremento del caseificio tenuto in Milano nel marzo 1874 presso la Regia Scuola Superiore di Agricoltura. A cura del Ministero di Agricoltura Industria e Commercio. Tipografia Editrice Lombarda, Milano, pp 183–189

Printed in the United States
By Bookmasters